STRUCTURAL ACOUSTICS

Deterministic and Random Phenomena

STRUCTURAL ACOUSTICS

Deterministic and Random Phenomena

JOSHUA E. GREENSPON

CRC Press
Taylor & Francis Group
Boca Raton London New York

CRC Press is an imprint of the
Taylor & Francis Group, an **informa** business

CRC Press
Taylor & Francis Group
6000 Broken Sound Parkway NW, Suite 300
Boca Raton, FL 33487-2742

First issued in paperback 2017

ISBN-13: 978-1-4398-3093-2 (hbk)
ISBN-13: 978-1-138-07562-7 (pbk)

Library of Congress Cataloging-in-Publication Data

Greenspon, Joshua E.
Structural acoustics : deterministic and random phenomena / Joshua E. Greenspon.
p. cm.
Includes bibliographical references and index.
ISBN 978-1-4398-3093-2 (hardcover : alk. paper)
1. Structural dynamics. I. Title.

TA654.G74 2011
620.2--dc23 2011019719

Visit the Taylor & Francis Web site at
http://www.taylorandfrancis.com

and the CRC Press Web site at
http://www.crcpress.com

Contents

Preface

This book is a primer for those familiar with the response of structures, elastic, and acoustic media. Most built-up structures are composed of beams, plates, and shells. Solution methods for computing the response of structures and media to static, dynamic, and random loading are provided.

The author has presented the material in two main sections containing a total of three parts. Section one contains Part 1 (Chapters 1 to 3), which give a review of the book and presents the fundamentals of acoustics and structural acoustics. Part 2 (Chapters 4 to 8) gives the basic analysis of beams, plates, built-up structures, cylindrical shells, and built-up variable media. Section two contains Part 3 (Chapters 9 through 14), which presents the analysis of and applications for random loading.

Methods for solving problems for complicated media that vary from point to point are presented. There are solutions for computing the response and the sound radiated from structures and elastic media that are in contact with fluid and are under static, dynamic, and random loading.

The author has derived simplified solutions to complicated problems. For example, there is a method for dividing a built-up shell structure into stiffened plate sections (Chapter 6), where the characteristics of the plate sections are determined in an earlier part of the book. A procedure is offered (Chapter 8) for computing the response of an elastic or viscoelastic media without resorting to a large computer program.

The fundamentals and formulas for random loading are presented. Sample cases are derived and curves are provided. The author discusses the fundamental aspects of simple structures and then expands to the more complicated cases with more involved loading.

Chapter 1 presents an overview of the entire book, including a description of some of the more important applications.

Chapter 2 presents the fundamentals of acoustics and structural acoustics.

Chapter 3 gives methods for computing added mass and radiation damping for plate-type structures.

Chapter 4 presents fundamentals of structures and analysis of beams, which are the simplest of the structures comprising a built-up system.

Chapter 5 discusses problems that are more complicated than beams (i.e., plates).

Chapter 6 shows how to use the theory of plates to analyze built-up structures.

Chapter 7 is devoted to details for analyzing cylindrical shells.

Chapter 8 presents the analysis of bodies with variable elastic, viscoelastic, and fluid media.

Chapter 9 starts random loading theory.

Chapter 10 is devoted to problems in random acoustics.

Chapter 11 extends the analysis to random problems in structures.

Chapter 12 shows how to obtain the acoustic field of a randomly loaded cylindrical shell.

Chapter 13 deals with applications of statistical acoustics to near field–far field problems.

Chapter 14 presents random response of scale models.

The author has attempted to show the reader how to solve a great many of the structural and acoustic problems that exist in today's technical world. The earlier part of the book shows how to solve beam and plate problems. The plate analysis is then applied to built-up shells in Chapter 6. How to solve problems of cylindrical shells is treated in detail in Chapters 7 and 12. Chapter 8 then treats infinite media made up of different materials. Deterministic problems with any type of loading are treated and systems with random loading are then discussed.

The five computer programs that are described in the text can be downloaded from the publisher's website: www.crcpress.com.

Acknowledgments

The author is very grateful to the late Marvin Lasky, formerly with the Office of Naval Research, for his guidance and encouragement when the author worked on structural acoustics under contract to the U.S. Navy. The author would also like to express his gratitude to Sonia Looban Greenspon, his wife, for editing and typing the manuscript, and to his daughter and son-in-law Ellen and Saul Singer for all of their assistance in the preparation of this book.

The Author

Dr. Greenspon received his bachelor's, master's, and doctoral degrees from The Johns Hopkins University, Baltimore, MD. While studying for his doctoral degree, he also taught applied mechanics for several years at the university. Upon completion of his course work, he was employed by the David Taylor Model Basin to do studies on ship vibrations. He completed the original research for his doctoral thesis in vibrations of thick cylindrical shells.

He subsequently worked for the Martin Company doing research in panel flutter. He later established a consultancy firm – JG Engineering Research Associates, and continued doing research under contract to the United States Army and Navy, in the areas of vibrations, sound radiation and structures subject to blast and fragmentation.

Dr. Greenspon was awarded the Silver Medal in Engineering Acoustics by The Acoustical Society of America for the solution of underwater radiation and scattering problems. He also served as an associate editor of the *Journal of the Acoustical Society of America* for 20 years. He has membership in the Acoustical Society of America, American Society of Mechanical Engineers, Society of Naval Architects and Marine Engineers, and the American Institute of Aeronautics and Astronautics.

Dr. Greenspon passed away in February 2011, a few months before the publication of this book.

Section 1

Deterministic Phenomena

1

General Overview of the Book

1.1 Introduction

This book presents simplified relations and computer programs for fundamental problems that one would encounter in practice. The structures that make up most of the complex systems can be divided into beams, plates, and shells. Analysis is presented for these structures as well as complex systems composed of a series of plates. A general program that has variable properties is also developed for elastic and viscoelastic media. Thus, solutions in acoustic and structural wave propagation of these media can be solved. All of this analysis provides the reader with methods for approaching a host of the problems of elastic and acoustic analysis of today's structures and surrounding media.

In Chapters 9–13, this analysis is employed to solve problems where the loading is random. The basic analysis of random systems is presented and the fundamental formulas are derived for random loading on an arbitrary system.

In each of the structures (beams, plates, and shells) considered in this book, the response is calculated for a concentrated harmonic load located at a single point. This response is called a Green's function for the structure under consideration. The response for any distributed load can be obtained by just dividing the distributed load in sections, calculating the response at each section, and then adding the contributions. For any time-wise loading, the loading is expanded into a Fourier series and the contributions from each of the frequency terms of the series are added up to obtain the desired response. Thus, the Green's function can be used for any spatial and time-wise load function and for input to the random loading problem. The acoustic field is computed by employing the methods outlined in the chapter on near field–far field methods by using the acceleration response.

Built-up systems such as buildings, ships, aircraft, and missiles are constructed of combinations of beams, plates, and shells. This book begins with methods of computing the response of these simple systems to point harmonic loads. Methods for obtaining the response to any other time or space loading using the point harmonic loading will be discussed.

A simple method for adding water loading to plate structures is shown in Chapter 3. Chapter 6 shows how to construct a system by using the plate elements. Methods for computing the acoustic field from the structural response are outlined in Chapter 13. For very complex systems, such as the ground, the ocean and its bottom, and the atmosphere in which the medium changes as a function of location, a system and computer program are given in Chapter 8. Chapters 9–13 are devoted to systems with random loading, where the functions given in the first part of the book can be applied.

In summary, the book first discusses the fundamentals of structural acoustics. It then shows how to solve problems of buildup systems, and finally it develops the fundamentals of random structural acoustics.

1.2 Background

Structural acoustics is the study of the response of structures that are coupled with a fluid and the resulting field generated in the fluid. Acoustic radiation is the field generated in the fluid due to the vibration of the structure. In contrast, acoustic scattering is the field generated when an acoustic wave is projected onto the structure. Random structural acoustics studies what occurs when the loading on the structure is random.

There are many books on vibrations of structures where the loading is a well-defined function of space and time. There is also much literature on random vibrations, where the loading is not a definite function of space and time but can be described by some statistical function (see the bibliographies in references 1–7).

There are very few references in the area of random loading on a structure surrounded by a fluid. One such book is *Structural Acoustics and Vibration*.[8] This book is a scholarly treatise. Papers that discuss this subject are in the reference section of the treatise. This is one of the areas on which this current book is focused. The author's intention here is to communicate the subject of random structural acoustics in a more practical format and to provide computer programs for the reader.

Although this book is entitled *Structural Acoustics*, the analysis also applies to nonacoustic media that are in contact with the structure. Thus, the overall theory can also be applied to geophysical phenomena.

The ideas developed here show that random response is dependent upon two major functions. The first is the Green's function of the structure, which is the response of the structure to a unit harmonic load. The second parameter is the cross spectrum of the loading. The resulting work is the cross spectrum of the response from which all of the important results, such as the spectrum of the response, the coherence, the mean square response, and others, are explained in the text.

Some of the important random loading functions are given in Nigam and Narayanan.[7] For aerospace vehicles, there are gust, boundary layer turbulence, and jet noise. For ground problems, there are earthquakes, wind loading, and offshore wave loading. For naval structures, there are turbulent boundary layer and random machinery forces, to mention a couple. A general method that has been used for these complex problems is statistical energy analysis, a procedure published in books by Lyon[9] and Dejong and Lyon.[10] The book by Lyon gives a complete history and develops the theory from the beginning.

The chief applications are associated with high-frequency response of systems, where many modes are involved. In contrast, the theory presented here is most useful for lower frequency motions, but it holds for any frequency. Using the computer capabilities that exist today, the complete theory of random response and sound radiation is well within the realm of practical application. The case of the cylindrical shell is worked out completely in this book.

The shell is assumed to be simply supported at its ends, but other cases such as fixed ends and free ends can be approximated by using the appropriate beam functions along the length and using an equivalent length. The cases of stiffened cylindrical shells, anisotropic and composite shells, isotropic shells, and sandwich shells are worked out. The analysis includes shells moving at a given speed. The far field radiation patterns are worked out for loads with given frequencies and for a turbulent boundary layer. Computer programs are included for these calculations.

1.3 Overview of Random Structural Acoustics

1.3.1 The Problems of Most Practical Interest

Random structural acoustics is the study of the acoustical field generated by random loading on a structure that is surrounded by a fluid. Some problems in which the acoustical field generated by a structure that are of interest are as follows:

- The sound radiation from ships and submarines from loading on the inside and outside
- The sound generated by jets and the corresponding response of the structures that are excited by the jet noise, which applies mostly to aircraft problems
- The sound generated by boundary layer turbulence in air or water, which applies to automotive problems, ships, submarines, and aircraft

In ships and submarines, the machinery produces forces on the hull that in turn produce sound radiation from the hull. This sound radiation is the factor that makes the ship or submarine identifiable. Considerable effort has to be used to quiet this sound radiation.

Jet engines are the power unit in most modem aircraft. Jets produce tremendous amounts of very high-level noise, which can excite vibrations in the structure of the aircraft as well as disturbance on the inside. The vibrations of the outside structure due to jet noise can cause fatigue failure in the aircraft structure; this can be disastrous to the safety of the aircraft. Jet noise must be reduced.

The sound generated by boundary layer noise can interfere with the sonar in ships and submarines and can be a disturbing influence in automotive structures.

1.4 Overview of Methods and Applications

1.4.1 Methods

For random loading problems, the general integral for the response is derived in Chapter 10. One can derive simpler relations from this general integral by introducing modal analysis. Several recent sources (see references 1–7) are devoted to random vibrations. Each of these sources has a complete bibliography of the subject dating from the 1950s up to the late 1990s. Reference 8 is one of the only references that provides a general overview of random structural acoustics in one of its chapters.

1.4.2 Green's Functions and Cross Spectrum

The basic connection between structural acoustics and random structural acoustics is the Green's function of the structure. The solution to the random structural acoustic problem involves two basic functions. The first is the Green's function, which is the response (acceleration, velocity, displacement, acoustic pressure) at one point due to a unit harmonic load at another point. The second function is the cross spectrum of the input. The final answer is in the form of cross spectra. Other statistical items of interest are derived from this cross spectrum of the output.

1.4.3 Modal Analysis

Chapter 11 contains the modal method of computing the response (motion and acoustic field). This method was developed by Powell[11,12] and extended by Greenspon[13] to compute the acoustic far field. Powell[12] shows how simplifications can be made to obtain the far field mean square pressure in terms of modes.

1.5 Comparison of Methods Used in the Recent Literature

Most of the recent literature uses the methods presented here that were developed in the 1950s, 1960s, and 1970s, as shown by the references in these recent books. Simplifications are made in some cases in order to make the problem tractable (see references 1–7 and the references contained therein for discussion of particular problems of interest). The reader is especially encouraged to examine Lalanne's book[5] to obtain the basic definitions associated with random processes.

The aim of this book is to present the basic equations for obtaining practical results of these random processes. It is intended to be a guide that can be employed when using finite element procedures to obtain results for complicated systems in random structural acoustics.

1.6 Applications

One of the most complete books on acoustic radiation is Hanish's treatise.[14] This book treats many of the important problems in acoustic radiation, including some of the practical random problems.

Structures can be divided into groups. The first group, using the most complicated case, is the three-dimensional body, which has different properties at each point. The equations for an elastic, viscoelastic or fluid body with variable properties are given in Chapter 8, which discusses approximate solutions for such a structure.

The next simplest structures are three-dimensional shells of various shapes but with relatively small surface thickness. Chapter 6 is devoted to this type of structure and Chapter 7 treats cylindrical shells, which are also in this group.

The next simplest case is the plate, which is treated in Chapter 5. The simplest case is the one-dimensional beam treated in Chapter 4.

An interesting phenomenon is that the three-dimensional body can be used to determine the solutions of beams, plates, and shells if certain assumptions that simplify the equations of the three-dimensional body are made. In the past, before small computers were available to study large systems, the theory of beams, plates, and shells was a separate entity and thousands of papers were published in hundreds of technical journals on the subject of the response of beams, plates, and shells.

In the modern world, where small computers can handle millions of elements, the three-dimensional problems can be solved. One type of solution for the dynamics of practical shell problems is contained in Chapter 6. Another type of solution for wave propagation is offered in Chapter 8 for the three-dimensional body with variable characteristics. The three-dimensional body can be solved with finite difference methods using the ideas given in Chapters 6 and 8.

It should be noted that the equations in Chapter 8 include fluid properties since the equations are for a variable parameter system. The fluid parts are obtained by letting the shear modulus, G, equal zero.

References

1. Elishakoff, I. 1983. *Probabilistic methods in the theory of structures.* New York: John Wiley & Sons.
2. Cederbaum, G., Elishakoff, I., Aboudi, J., and Librescu, L. 1992. *Random vibration and reliability of composite structures.* Lancaster, PA: Technomic Publishing Co.
3. Yang, C. Y. 1986. *Random vibration of structures.* New York: John Wiley & Sons.
4. Elishakoff, I., Lin, Y. K., and Zhu, L. P. 1994. *Probabilistic and convex modeling of acoustically excited structures.* New York: Elsevier.
5. Lalanne, C. 2002. *Mechanical vibration & shock—Random vibration,* vol. III. London: Hermes Science Publications.
6. Cook, N. J. 1985. *The designer's guide to wind loading of building structures.* Cambridge, England: Cambridge University Press.
7. Nigam, N. C., and Narayanan, S. 1994. *Application of random vibrations.* New York: Springer–Verlag.
8. Ohayon, R., and Soize, C. 1998. *Structural acoustics and vibration.* New York: Academic Press.
9. Lyon, R. H. 1975. *Statistical energy analysis of dynamical systems: Theory and applications.* Cambridge, MA: MIT Press.
10. Lyon, R. H., and DeJong, R. G. 1995. *Theory and application of statistical energy analysis.* Boston, MA: Butterworth-Heinemann.
11. Powell, A. 1958. On the response of structures to random pressures and to jet noise in particular. In *Random vibration,* ed. S. H. Crandall, 187. New York: John Wiley & Sons.
12. Powell, A. 1957. On structural vibration excited by random pressures with reference to structural fatigue and boundary layer noise. Douglas Aircraft Co., report no. SM 22795.
13. Greenspon, J. E. 1967. Far-field radiation from randomly vibrating structures. *Journal of Acoustical Society of America* 41 (5): 1201–1207.
14. Hanish, S. 1981. *A treatise on acoustic radiation.* Washington, DC: Naval Research Laboratory.

2

*Fundamentals of Acoustics and Structural Acoustics**

2.1 Introduction

A unified treatment of the principles of linear acoustics must begin with the well-known phenomenon of single-frequency acoustics. A second essential topic is random linear acoustics, a relatively new field, which is given a tutorial treatment later in this book. The objective is to present the elementary principles of linear acoustics and then to use straightforward mathematical development to describe some advanced concepts.

Section 2.2 gives a physical description of phenomena in acoustics. Section 2.3 starts with the difference between linear and nonlinear acoustics and leads to the derivation of the basic wave equation of linear acoustics. Section 2.4 addresses the intensity and energy of the one-dimensional harmonic wave. Section 2.5 discusses the fundamentals of normal-mode and ray acoustics, which is used extensively in studies of underwater sound propagation. In Section 2.6, details are given on sound propagation as it is affected by barriers and obstacles. Sections 2.7 and 2.8 deal with waves in confined spaces; sound radiation, with methods of solution to determine the sound radiated by structures; and the coupling of sound with its surroundings. All of these areas are discussed in references 1–12.

2.2 Physical Phenomena in Linear Acoustics

2.2.1 Sound Propagation in Air, Water, and Solids

Many practical problems are associated with the propagation of sound waves in air or water. Sound does not propagate in free space. It must have a dense

* This chapter was published in *Encyclopedia of Physical Science and Technology*, vol. 1, Greenspon, J. E., 114–152. Copyright Elsevier (1992).

medium in which to propagate. Thus, for example, when a sound wave is produced by a voice, the air particles in front of the mouth are vibrated, and this vibration, in turn, produces a disturbance in the adjacent air particles, and so on.

If the wave travels in the same direction in which the particles are being moved, it is called a longitudinal wave. This same phenomenon occurs whether the medium is air, water, or a solid. If the wave is moving perpendicularly to the moving particles, it is called a transverse wave.

The rate at which a sound wave thins out, or attenuates, depends to a large extent on the medium through which it is propagating. For example, sound attenuates more rapidly in air than in water, which is the reason that sonar is used more extensively underwater than in air. Conversely, radar (electromagnetic energy) attenuates much less in air than in water, so it is more useful in air as a communication tool.

Sound waves travel in solid or fluid materials by elastic deformation of the material, which is called an elastic wave. In air (below a frequency of 20 kHz) and in water, a sound wave travels at constant speed without its shape being distorted. In solid material, the velocity of the wave changes, and the disturbance changes shape as it travels. This phenomenon in solids is called dispersion. Air and water are, for the most part, nondispersive media, whereas most solids are dispersive media.

2.2.2 Reflection, Refraction, Diffraction, Interference, and Scattering

Sound propagates undisturbed in a nondispersive medium until it reaches some obstacle. The obstacle, which can be a density change in the medium or a physical object, distorts the sound wave in various ways. It is interesting to note that sound and light have many propagation characteristics in common: The phenomena of reflection, refraction, diffraction, interference, and scattering for sound are very similar to the phenomena for light.

2.2.2.1 Reflection

When sound impinges on a rigid or elastic obstacle, part of it bounces off the obstacle, a characteristic that is called reflection. The reflection of sound back toward its source is called an echo. Echoes are used in sonar to locate objects underwater. Most people have experienced echoes in air by shouting in an empty hall and hearing their words repeated as the sound bounces off the walls.

2.2.2.2 Refraction and Transmission

Refraction is the change of direction of a wave when it travels from a medium in which it has one velocity to a medium in which it has a different velocity. Refraction of sound occurs in the ocean because the temperature or the water changes with depth, which causes the velocity of sound also to change with depth. For simple ocean models, the layers of water at different temperatures act as though they are layers of different media. The following example explains refraction: Imagine a sound wave that is constant over a plane (i.e., a plane wave) in a given medium and a line drawn perpendicularly to this plane (i.e., the normal to the plane), which indicates the travel direction of the wave. When the wave travels to a different medium, the normal bends, thus changing the direction of the sound wave. This normal line is called a ray and is discussed later in the discussion of ray acoustics.

When a sound wave impinges on a plate, part of the wave reflects and part goes through the plate. The part that goes through the plate is the transmitted wave. Reflection and transmission are related phenomena that are used extensively to describe the characteristics of sound baffles and absorbers.

2.2.2.3 *Diffraction*

Diffraction is associated with the bending of sound waves around or over barriers. A sound wave can often be heard on the other side of a barrier even if the listener cannot see the source of the sound. However, the barrier projects a shadow, called the shadow zone, within which the sound cannot be heard. This phenomenon is similar to that of a light that is blocked by a barrier.

2.2.2.4 *Interference*

Interference is the phenomenon that occurs when two sound waves converge. In linear acoustics, the sound waves can be superimposed. When this occurs, the waves interfere with each other, and the resultant sound is the sum of the two waves, taking into consideration the magnitude and the phase of each wave.

2.2.2.5 *Scattering*

Sound scattering is related closely to reflection and transmission. It is the phenomenon that occurs when a sound wave envelops an obstacle and breaks up, producing a sound pattern around the obstacle. The sound travels in all directions around the obstacle. The sound that travels back toward the source is called the backscattered sound, and the sound that travels away from the source is known as the forward scattered field.

2.2.3 Standing Waves, Propagating Waves, and Reverberation

When a sound wave travels freely in a medium without obstacles, it continues to propagate unless it is attenuated by some characteristic of the medium, such as absorption. When sound waves propagate in an enclosed space, they reflect from the walls of the enclosure and travel in a different direction until they hit another wall. In a regular enclosure, such as a rectangular room, the waves reflect back and forth between the sound source and the wall, setting up a constant wave pattern that no longer shows the characteristics of a traveling wave. This wave pattern, called a standing wave, results from the superposition of two traveling waves propagating in opposite directions.

The standing wave pattern exists as long as the source continues to emit sound waves. The continuous rebounding of the sound waves causes a reverberant field to be set up in the enclosure. If the walls of the enclosure are absorbent, the reverberant field is decreased. If the sound source stops emitting the waves, the reverberant standing wave field dies out because of the absorptive character of the walls. The time it takes for the reverberant field to decay is sometimes called the time constant of the room.

2.2.4 Sound Radiation

The interaction of a vibrating structure with a medium produces disturbances in the medium that propagate out from the structure. The sound field set up by these propagating disturbances is known as the sound radiation field. Whenever there is a disturbance in a sound medium, the waves propagate out from the disturbance, forming a radiation field.

2.2.5 Coupling and Interaction between Structures and the Surrounding Medium

A structure vibrating in air produces sound waves, which propagate into the air. If the same vibrating structure is put into a vacuum, no sound is produced. However, whether the vibrating body is in a vacuum or air makes little difference in the vibration patterns, and the reaction of the structure to the medium is small. If the same vibrating body is placed into water, the high density of water compared with air produces marked changes in the vibration and consequent radiation from the structure. The water or any heavy liquid produces two main effects on the structure. The first is an added mass effect. The second is a damping effect known as radiation damping. The same type of phenomenon also occurs in air, but to a much smaller degree, unless the body is traveling at high speed. The coupling phenomenon in air at these speeds is associated with flutter.

2.2.6 Deterministic versus Random Linear Acoustics

When the vibrations are not single frequency but are random, new concepts must be introduced. Instead of dealing with ordinary parameters such as pressure and velocity, it is necessary to use statistical concepts such as auto- and cross correlation of pressure in the time domain and auto- and cross spectra of pressure in the frequency domain. Frequency is a continuous variable in random systems, as opposed to a discrete variable in single-frequency systems. In some acoustic problems, there is randomness in both space and time. Thus, statistical concepts have to be applied to both time and spatial variables.

2.3 Basic Assumptions and Equations in Linear Acoustics

2.3.1 Linear versus Nonlinear Acoustics

The basic difference between linear and nonlinear acoustics is determined by the amplitude of the sound. The amplitude is dependent on a parameter, called the condensation, that describes how much the medium is compressed as the sound wave moves. When the condensation reaches certain levels, the sound becomes nonlinear. The major difference between linear and nonlinear acoustics can best be understood by deriving the one-dimensional wave equation for sound waves and studying the parameters involved in the derivation. Consider a plane sound wave traveling down a tube, as shown in Figure 2.1.

Let the cross-sectional area of the tube be A and let ξ be the particle displacement along the x axis from the equilibrium position. Applying the principle of conservation of mass to the volume $A\,dx$ before and after it is displaced, the following equation is obtained:

$$\rho A\,dx(1+\partial\xi/\partial x)=\rho_0 A\,dx \tag{2.1}$$

The mass of the element before the disturbance arrives is $\rho_0 A\,dx$ where ρ_0 is the original density of the medium. The mass of this element as the disturbance passes is changed to

$$\rho A\,dx(1+\partial\xi/\partial x) \tag{2.1a}$$

where ρ is the new density of the disturbed medium. This disturbed density p can be defined in terms of the original density ρ_0 by the following relation:

$$\rho=\rho_0(1+S) \tag{2.2}$$

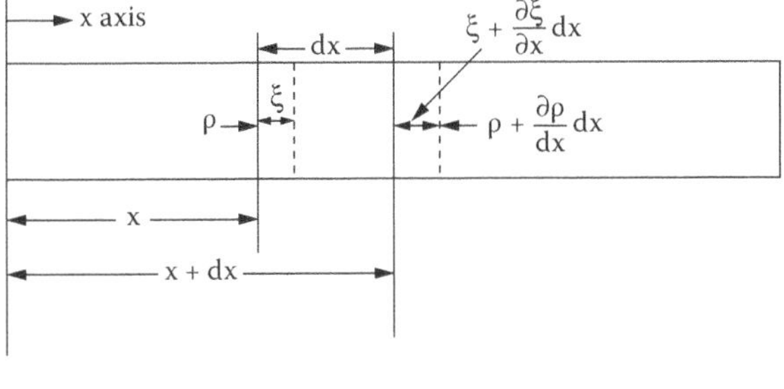

FIGURE 2.1
Propagation of a plane one-dimensional sound wave. A = cross-sectional area of tube; ξ = particle displacement along x-axis; p = acoustic pressure.

where S is called the condensation. By substituting Equation 2.2 into Equation 2.1a, we obtain

$$(1+S)(1+\partial\xi/\partial x)=1 \tag{2.3}$$

If p is the sound pressure at x, then $p + \partial p/\partial x\, dx$ is the sound pressure at $x + dx$ (by expanding p into a Taylor series in x and neglecting higher order terms in dx). Applying Newton's law to the differential element, we find that

$$-\frac{\partial p}{\partial x}=\rho_o\frac{\partial^2\xi}{\partial t^2} \tag{2.4}$$

If it is assumed that the process of sound propagation is adiabatic (i.e., there is no change of heat during the process), then the pressure and density are related by the following equation:

$$\frac{P}{p_o}=\left(\frac{\rho}{\rho_o}\right)^{\gamma} \tag{2.5}$$

where P = total pressure = $p + p_o$, p is the disturbance sound pressure, and γ is the adiabatic constant, which has a value of about 1.4 for air. Using Equations 2.2 and 2.3 gives

$$\rho=\frac{\rho_o}{1+\partial\xi/\partial x} \tag{2.5a}$$

Thus,

$$\frac{\partial p}{\partial x}=\frac{\partial P}{\partial x}=-\gamma p_o\left(1+\frac{\partial\xi}{\partial x}\right)^{-1-\gamma}\frac{\partial^2\xi}{\partial x^2} \tag{2.6}$$

Substituting into Equation 2.4 gives

$$\gamma p_o\frac{\partial^2\xi/\partial x^2}{(1+\partial\xi/\partial x)^{1+\gamma}}=\rho_o\frac{\partial^2\xi}{\partial t^2} \tag{2.7}$$

or, finally,

$$c^2\frac{\partial^2\xi/\partial x^2}{(1+\partial\xi/\partial x)^{1+\gamma}}=\frac{\partial^2\xi}{\partial t^2} \tag{2.8}$$

where $c^2 = p_o\gamma/\rho_o$ (c is the sound speed in the medium). If $\partial\xi/\partial x$ is small compared with 1, then Equation 2.3 gives

$$S = -\frac{\partial \xi}{\partial x} \tag{2.9}$$

and gives

$$c^2 \frac{\partial^2 \xi}{\partial x^2} = \frac{\partial^2 \xi}{\partial t^2} \tag{2.10}$$

Thus,

$$\xi = f_1(x - ct) + f_2(x - ct) \tag{2.10a}$$

Equations 2.9 and 2.10 are the linear acoustic approximations. The first term in Equation 2.10a is an undistorted traveling wave that is moving at speed c in the $+x$ direction, and the second term is an undistorted traveling wave moving with speed c in the $-x$ direction.

Condensation values S of the order of 1 are characteristic of sound waves with amplitudes approaching 200 dB rel. 0.0002 dyne/cm^2. The threshold of pain is about 130 dB rel. 0.0002 dyne/cm^2. This is the sound pressure level that results in a feeling of pain in the ear. The condensation value for this pressure is about $S = 0.0001$. For a condensation value $S = 0.1$, we are in the nonlinear region. This condensation value corresponds to a sound pressure level of about 177 dB rel. 0.0002 dyne/cm^2. All the ordinary sounds that we hear, such as speech and music (even very loud music), are usually well below 120 dB rel. 0.0002 dyne/cm^2. A person who is very close to an explosion or is exposed to sonar transducer sounds underwater would suffer permanent damage to his hearing because the sounds are usually well above 130 dB rel. 0.0002 dyne/cm^2.

2.3.2 Derivation of Basic Equations

It is now necessary to derive the general three-dimensional acoustic wave equations. In a previous section, the one-dimensional wave equation was derived for both the linear and nonlinear cases. If there is a fluid particle in the medium with coordinates x, y, and z, the fluid particle can move in three dimensions. Let the displacement vector of the particle be b having components ξ, η, and ζ as shown in Figure 2.2.

The velocity vector **q** is

$$\mathbf{q} = \partial \mathbf{b}/\partial t \tag{2.11}$$

Let this velocity vector have components u, υ, and w, where

$$u = \partial \xi/\partial t \qquad \upsilon = \partial \eta/\partial t \qquad w = \partial \zeta/\partial t \tag{2.12}$$

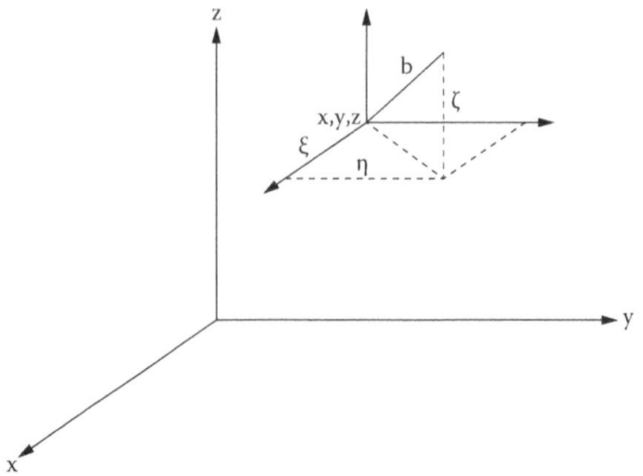

FIGURE 2.2
The fluid particle, x, y, and z rectangular coordinates of fluid particle; b = displacement vector of the particle (components of b are ξ, η, and ζ).

As the sound wave passes an element of volume $V = dx\,dy\,dz$, the element changes volume because of the displacement ξ, η, and ζ. The change in length of the element in the x, y, and z directions, respectively, is shown in Figure 2.2 above.

So, the new volume is $V + \Delta V$, where

$$\begin{aligned} V + \Delta V = dx\left(1 + \frac{\partial \xi}{\partial x}\right) \\ dy\left(1 + \frac{\partial \eta}{\partial y}\right) dz\left(1 + \frac{\partial \rho}{\partial z}\right) \end{aligned} \tag{2.13}$$

The density of the medium before displacement is ρ_o and the density during displacement is $\rho_o(1 + S)$, as in the one dimensional case developed in the last section. Applying the principle of conservation of mass to the element before and after displacement, we find that

$$(1 + S)(1 + \partial \xi/\partial x)(1 + \partial \eta/\partial y)(1 + \partial \zeta/\partial z) = 1 \tag{2.14}$$

Now we make the linear acoustic approximation that $\partial \xi/\partial x$, $\partial \eta/\partial y$, and $\partial \zeta/\partial z$ are small compared with 1. So, Equation 2.14 becomes the counterpart of Equation 2.9 in one dimension:

$$S = -(\partial \xi/\partial x + \partial \eta/\partial y + \partial \zeta/\partial z) \tag{2.15}$$

This equation is called the equation of continuity for linear acoustics.

The equations of motion for the element *dx dy dz* are merely three equations in the three coordinate directions that parallel the one-dimensional Equation 2.4; thus, the three equations of motion are

$$-\frac{\partial p}{\partial x} = \rho_0 \frac{\partial^2 \xi}{\partial t^2} \qquad -\frac{\partial p}{\partial y} = \rho_0 \frac{\partial^2 y}{\partial t^2}$$
$$-\frac{\partial p}{\partial z} = \rho_0 \frac{\partial^2 \rho}{\partial t^2} \tag{2.16}$$

If one differentiates the first of these equations with respect to x, the second with respect to y, and the third with respect to z and adds them, and then, letting

$$\nabla^2 = \partial^2/\partial x^2 + \partial^2/\partial y^2 + \partial^2/\partial z^2$$

one obtains

$$\nabla^2 p = \rho_0 \frac{\partial^2 S}{\partial t^2} \tag{2.17}$$

Now we introduce the adiabatic assumption in Equation 2.17—that is,

$$\frac{P}{\rho_0} = \left(\frac{\rho}{\rho_0}\right)^{\gamma} \tag{2.18}$$

where P = total pressure = $p + p_o$ and p is the sound pressure due to the disturbance. Since

$$\rho = \rho_o(1+S),$$
$$P/p_o = (1+S)^{\gamma} \tag{2.19}$$

For small S, the binomial theorem applied to Equation 2.19 gives

$$\frac{P - p_0}{\rho_0} = Sc^2 \tag{2.20}$$

(c being the adiabatic sound velocity, as discussed for the one-dimensional case). Thus,

$$p = \rho_o Sc^2 \tag{2.20a}$$

Substituting into Equation 2.17, we obtain

$$c^2\nabla^2 p = \partial^2 p/\partial t^2 \tag{2.21}$$

2.4 Intensity and Energy

The one-dimensional equation for plane waves is given by Equation 2.10. The displacement for a harmonic wave can be written as

$$\xi = Ae^{i(\omega t + kx)} \tag{2.21a}$$

The pressure is given by Equation 2.20a—that is, $\rho_o C_o^2 S$, where $S = \partial\xi/\partial x$ for the one-dimensional wave. Then,

$$p = \rho_o C_o^2 \frac{\partial \xi}{\partial x} \tag{2.21b}$$

The velocity is given by $u = \partial\xi/\partial t$, so, for one-dimensional harmonic waves, $p = \rho_o c_o^2(ik)$ and $u = i\omega\xi$, $k = \omega/c_o$. Thus, $p = \rho_o c_o u$. The intensity is defined as the power flow per unit area for the rate at which energy is transmitted per unit area. Thus, $I = p\xi$. The energy per unit area is the work done on the medium by the pressure in going through displacement ξ; that is, $E_f = p\xi$, by the above, $I = p^2/\rho_o c_o$.

2.5 Free Sound Propagation

2.5.1 Ray Acoustics

Characteristics of sound waves can be studied by the same theory regardless of whether the sound is propagating in air or water. The simplest of the sound-propagation theories is known as ray acoustics. A sound ray is a line drawn normal to the wave front of the sound wave as the sound travels. In Section 2.2.2, refraction was described as the bending of sound waves when going from one medium to another. When a medium such as air or water has a temperature gradient, then it can be thought of as having layers that act as different media. The objective of this theory or any other transmission theory is to describe the sound field at a distance from the source.

There are two main equations of ray theory. The first is Snell's law of refraction, which states that

$$\frac{V_1}{\cos\theta_1} = \frac{V_2}{\cos\theta_2} = \frac{V_3}{\cos\theta_3} = \cdots = \frac{V_n}{\cos\theta_n} \tag{2.22}$$

where V_1, V_2,...V_n are the velocities of sound through the various layers of the medium, which are at different temperatures, as shown in Figure 2.3.

The second relation is that the power flow remains constant along a ray tube (i.e., there is conservation of energy along a ray tube). A ray tube is a closed surface formed by adjacent rays, as shown in Figure 2.4. If the power flow remains constant along a ray tube, then

$$\frac{p_1^2 A_1}{\rho_o c_1} = \frac{p_2^2 A_2}{\rho_o c_2} = \cdots \frac{p_n^2 A_n}{\rho_o c_n} \tag{2.23}$$

where
p refers to the sound pressure
A refers to the cross-sectional area of the ray tube
ρ_o refers to the mass density of the medium
C refers to the sound velocity in the medium

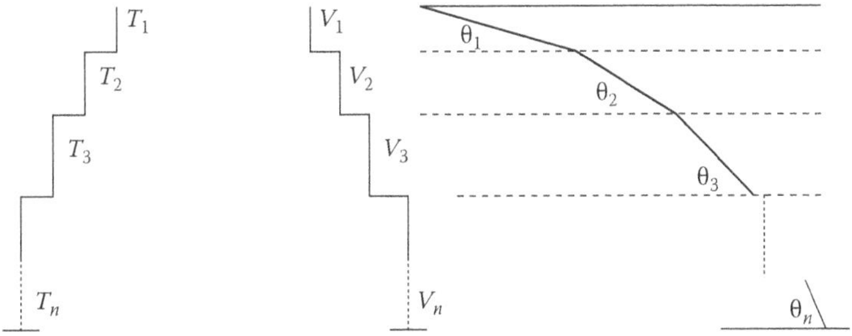

FIGURE 2.3
Temperatures, velocities, and refraction angles of the sound. T_1, T_2...T_n = temperatures of the n layers of the model medium; V_1, V_2...V_n = sound velocities in the n layers of the model medium.

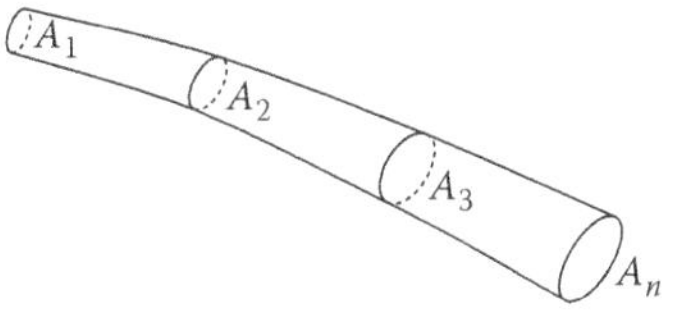

FIGURE 2.4
Ray tube. A_1, A_2...A_n = cross section area of the ray tube at the n stations along the tube.

But $p^2/\rho_0 c = I$, the sound intensity. Thus,

$$I_1 A_1 = I_2 A_2 = \cdots I_n A_n \tag{2.24}$$

The intensity can therefore be found at any point if the initial intensity I_1 and the areas of the ray tube $A_1,\ldots A_n$ are known. The ray tube and the consequent areas can be determined by tracing the rays. The tracing is done by using Snell's law (Equation 2.22). The velocities of propagation $V_1, V_2,\ldots V_n$ are determined from the temperature and salinity of the layer. One such equation for sound velocity is

$$\begin{aligned} V &= 1449 + 4.6T - 0.055T^2 + 0.0003T^3 \\ &+ (1.39 - 0.012T)(s - 35) + .017d \end{aligned} \tag{2.25}$$

where
V is the velocity of sound in meters per second
T is the temperature in degrees centigrade
s is the salinity in parts per thousand
d is the depth in meters

The smaller the ray tubes are—that is, the closer together the rays are—the more accurate are the results.

Simple ray acoustics theory is good only at high frequencies (usually in the kilohertz region). For low frequencies (e.g., less than 100 Hz), another theory, the normal mode theory, has to be used to compute transmission characteristics.

2.5.2 Normal Mode Theory

The normal mode theory consists of forming a solution of the acoustic wave equation that satisfies specific boundary conditions. Consider the sound velocity $C(z)$ as a function of the depth z, and let h be the depth of the medium. The medium is bounded by a free surface at $z = 0$ and a rigid bottom at $z = h$.

Let a point source be located at $z = z_1$, $r = 0$ (in cylindrical coordinates) as shown in Figure 2.5. The pressure p is given by the Helmholtz equation:

$$\begin{aligned} &\frac{\partial^2 p}{\partial r^2} + \frac{1}{r}\frac{\partial p}{\partial r} + \frac{\partial^2 p}{\partial z^2} + k^2(z)p = -\frac{2}{r}\delta(z - z_1)\delta(r) \\ &k^2(z) = \frac{\omega^2}{c(z)^2} \end{aligned} \tag{2.26}$$

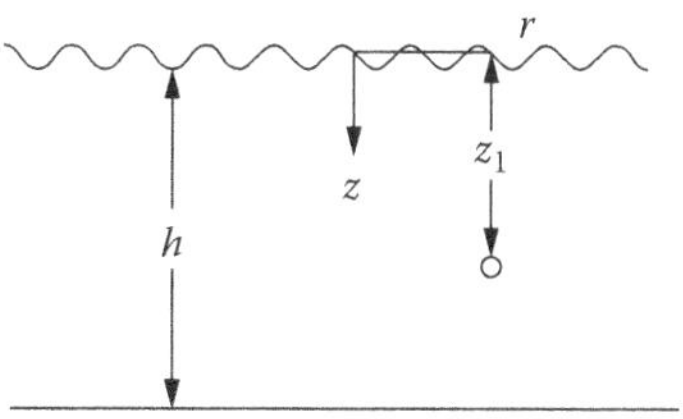

FIGURE 2.5
Geometry for normal mode propagation, $r; z$ = cylindrical coordinates of a point in the medium; h = depth of medium; z_1 = z coordinate of the source.

The δ functions describe the point source. The boundary conditions are

$$p(r,o) = 0 \qquad \text{(free surface)}$$

$$\frac{\partial p}{\partial z}(r,h) = 0 \qquad \text{(rigid bottom)} \tag{2.27}$$

Equations 2.26 and 2.27 essentially constitute all of the physics of the solution. The rest is mathematics. The solution of the homogeneous form of Equation 2.26 is first found by separation of variables. Since the wave has to be an outgoing wave, this solution is

$$p(r,z) = H_0^{(1)}(\xi r)\psi(z,\xi) \tag{2.28}$$

where $H_0^{(1)}$ is the Hankel function of the first kind of order zero. The function $\psi(z, \xi)$ satisfies the equation

$$\frac{d^2\psi}{dz^2} + [k^2(z) - \xi^2]\psi = 0 \tag{2.29}$$

with boundary conditions

$$\psi(o,\xi) = 0 \qquad \frac{d\psi(h,\xi)}{dz} = 0 \tag{2.30}$$

Since Equation 2.29 is a second-order linear differential equation, let the two linearly independent solutions be $\psi_1(z, \xi)$ and $\psi_2(z, \xi)$. Thus, the complete solution is

$$\psi(z,\xi) = B_1\psi_1(z,\xi) + B_2\psi_2(z,\xi) \tag{2.31}$$

where B_1 and B_2 are constants.

Substitution of Equation 2.31 into Equation 2.30 leads to an equation from which the allowable values of ξ (the eigenvalues) can be obtained—that is,

$$\psi_1(o,\xi)\frac{d\psi_2(h,\xi)}{dz} - \psi_2(o,\xi)\frac{d\psi_1(h,\xi)}{dz} = 0 \tag{2.32}$$

The nth root of this equation is called ξ_n. The ratio of the constants B_1 and B_2 is

$$\frac{B_1}{B_2} = -\frac{\psi_2(o,\xi_n)}{\psi_1(o,\xi_n)} \tag{2.33}$$

The $H_o^{(1)}(\xi_n r)\psi(z,\xi_n)$ are known as the normal mode functions, and the solution of the original inhomogeneous equation can be expanded in terms of these normal mode functions as follows:

$$p(r,z) = \sum_n A_n H_0^{(1)}(\xi_n r)\psi(z,\xi_n) \tag{2.34}$$

with unknown coefficients A_n, which will be determined next. Substituting Equation 2.34 into Equation 2.26 and employing the relation for the Hankel function

$$\left(\frac{d^2}{dr^2} + \frac{1}{r}\frac{d}{dr} + \xi_n^2\right)H_0^{(1)}(\xi_n r) = \frac{2i}{\pi r}\delta(r) \tag{2.35}$$

leads to

$$\sum_n A_n\psi_n(z) = i\pi\delta(z - z_1) \tag{2.36}$$

Next, one must multiply Equation 2.36 by $\psi_m(z)$ and integrate over the depth 0 to h. Using the orthogonality of mode assumption, which states that

$$\int_0^h \psi_n(z)\psi_m(z)dz = 0 \quad \text{if } m \neq n \tag{2.37}$$

we find that

$$A_n = \frac{i\pi\psi_n(z_1)}{\int_0^h \psi_n^2(z)dz} \tag{2.38}$$

So,

$$p(r,z) = \pi i \sum_n \psi_n(z_1)\psi_n(z) \frac{H_0^{(1)}(\xi_n r)}{\int_0^h \psi_n^2(z)dz} \tag{2.39}$$

If the medium consists of a single layer with constant velocity, c_o, it is found that

$$\begin{aligned} \psi_n(z) &= \cosh b_n z \qquad \xi_n = \sqrt{b_n^2 + k^2} \\ b_n &= i(n + \tfrac{1}{2})\pi/h \qquad k = \omega/c_o \end{aligned} \tag{2.40}$$

2.5.3 Underwater Sound Propagation

Ray theory and normal mode theory are used extensively in studying the transmission of sound in the ocean. At frequencies below 100 Hz, the normal mode theory is necessary. Two types of sonar are used underwater: active and passive. Active sonar produces sound waves that are sent out to locate objects by receiving echoes. Passive sonar listens for sounds. Since the sound rays are bent by refraction, there are shadow zones in which the sound does not travel. Thus, a submarine located in a shadow zone has very little chance of being detected by sonar.

Since sound is the principal means of detection underwater, there has been extensive research in various aspects of this field. The research has been divided essentially into three areas: generation, propagation, and signal processing. Generation deals with the mechanisms of producing the sound, propagation deals with the transmission from the source to the receiver, and signal processing deals with analyzing the signal to extract information.

2.5.4 Atmospheric Sound Propagation

It has been shown that large amplitude sounds such as sonic booms from supersonic aircraft can be detected at very low frequencies (called infrasonic frequencies) at distances above 100 km from the source. In particular, the Concorde sonic boom was studied at about 300 km and signals of about 0.6 N/m^2 were received at frequencies of the order of 0.4 Hz. The same phenomenon occurs for thunder and explosions on the ground. The same principles hold in the atmosphere as in water for the bending of rays in areas of changing temperature. Sound energy is not used for communication in air because of the large attenuation of higher frequency sound in air, as opposed to water. For example, considering the various mechanisms of absorption in the atmosphere, the total attenuation is about 24 dB per kiloyard at 100 Hz, whereas the sound attenuates underwater at 100 Hz at about 0.001 dB per kiloyard.

2.6 Sound Propagation with Obstacles

2.6.1 Refraction

Refraction is the bending of sound waves. The transmission of sound through the water with various temperature layers and the application of Snell's law have been treated in this chapter. Transmission of sound through water is probably the most extensive practical application of refraction in acoustics.

2.6.2 Reflection and Transmission

There are many practical problems related to reflection and transmission of sound waves. One example is used here to acquaint the reader with the concepts involved in the reflection and transmission of sound.

Consider a sound wave coming from one medium and hitting another, as shown in Figure 2.6. What happens in layered media, such as the temperature layers described in connection with ray acoustics and underwater sound, can now be noted. When ray acoustics and underwater sound were discussed, only refraction and transmission were described. The entire process for one transition layer can now be explained.

The mass density and sound velocity in the upper medium is ρ, c and in the lower medium is ρ_1, c_1. The pressure in the incident wave, p_{inc}, can be written as

$$p_{inc} = p_o e^{ik(x\sin\theta - z\cos\theta)} \quad k = \omega/c \tag{2.41}$$

(i.e., assuming a plane wave front).

From Snell's law, it is found that the refracted wave angle θ_1 is determined by the relation

$$\frac{\sin\theta}{\sin\theta_1} = \frac{c}{c_1} \tag{2.42}$$

There is also a reflected wave that goes off at angle θ, as shown in Figure 2.6. The magnitude of the reflected wave is not known, but since its direction θ is known, it can be written in the form

$$p_{refl} = Ve^{ik(x\sin\theta + z\cos\theta)} \tag{2.43}$$

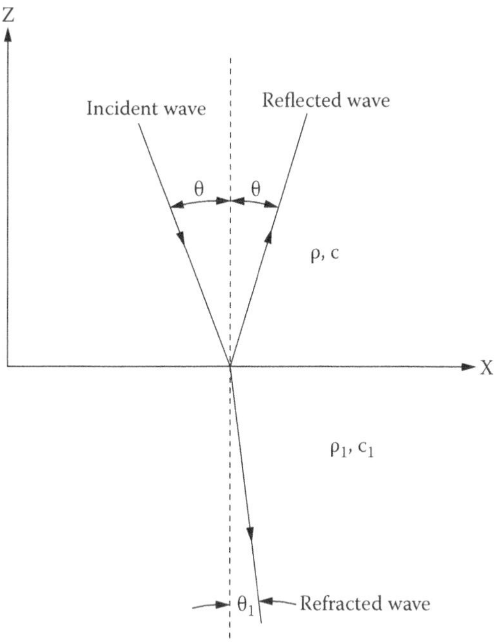

FIGURE 2.6
Reflection and transmission from an interface. ρ, c-mass density and sound velocity in upper medium; ρ_1, c_1 = mass density and sound velocity in lower medium; θ = angle of incidence and reflection; θ_1 = angle of refraction.

where V is the reflection coefficient. Similarly, the refracted wave can be written in the form

$$p_{\text{refrac}} = Wp_o e^{ik_1(x\sin\theta_1 - z\cos\theta_1)} \tag{2.44}$$

where $k_1 = \omega/c_1$ and W is the transmission coefficient. The boundary conditions at $z = 0$ are

$$p_{\text{upper}} = p_{\text{lower}} \quad \text{(acoustic pressure is continuous across the boundary)} \tag{2.45}$$

$$(\upsilon_z)_{\text{upper}} = (\upsilon_z)_{\text{lower}} \quad \text{(particle velocity normal to boundary is continuous across the boundary)}$$

The velocity is related to the pressure by the expression

$$\frac{\partial p}{\partial z} = \rho \frac{\partial \upsilon_z}{\partial t} \tag{2.46}$$

For harmonic motion, $\upsilon \sim e^{i\omega t}$, so

$$\partial p/\partial z = i\omega\rho\upsilon_z \tag{2.47}$$

The second boundary condition at $z = 0$ is therefore

$$\frac{1}{\rho}\frac{\partial p_{\text{upper}}}{\partial z} = \frac{1}{\rho_1}\frac{\partial p_{\text{lower}}}{\partial z} \tag{2.48}$$

The total field in the upper medium consists of the reflected and incident waves combined, so

$$\begin{aligned} p_{\text{upper}} &= p_{\text{inc}} + p_{\text{refl}} \\ &= p_o e^{ikx\sin\theta}(e^{-ikz\cos\theta} + Ve^{ikz\cos\theta}) \end{aligned} \tag{2.49}$$

Substituting into the boundary conditions, we find that

$$p_o e^{ikx\sin\theta}(1+V) = Wp_o e^{ik_1x\sin\theta_1}$$

Therefore,

$$1+V = We^{ik_1x\sin\theta_1 - ikx\sin\theta} \tag{2.50}$$

Since $1 + V$ is independent of x, then $e^{ik_1x\sin\theta_1 - ikx\sin\theta}$ must also be independent of x. Thus,

$$k_1\sin\theta_1 = k\sin\theta \tag{2.51}$$

which is Snell's law. Thus, the first boundary condition leads to Snell's law. The second boundary condition leads to the following equation:

$$\begin{aligned} &\frac{1}{\rho}p_o e^{ikx\sin\theta}(-ik\cos\theta + Vik\cos\theta) \\ &= \frac{1}{\rho_1}Wp_o e^{ik_1x\sin\theta_1}(-ik_1\cos\theta_1) \end{aligned} \tag{2.52}$$

Substituting Equation 2.51 into Equation 2.50 gives

$$1 + V = W \tag{2.53}$$

and substituting Equation 2.53 into Equation 2.52 gives

$$\frac{1}{\rho}e^{ikx\sin\theta}(-ik\cos\theta + Vik\cos\theta) = \frac{1}{\rho_1}(1+V)e^{ik_1x\sin\theta_1}(-ik_1\cos\theta_1) \tag{2.54}$$

or

$$\frac{\rho_1}{\rho}(-ik\cos\theta + Vik\cos\theta) = (1+V)(-ik_1\cos\theta_1)$$

Thus,

$$V = \frac{(\rho_1/\rho)k\cos\theta - k_1\cos\theta_1}{(\rho_1/\rho)k\cos\theta + k_1\cos\theta_1} \tag{2.55}$$

$$= \frac{\frac{\rho_1}{\rho}\cos\theta - \frac{k_1}{k}\cos\theta_1}{\frac{\rho_1}{\rho}\cos\theta + \frac{k_1}{k}\cos\theta_1} \qquad \frac{k_1}{k} = \frac{c}{c_1} \tag{2.56}$$

Equations 2.51, 2.53, and 2.56 give the unknowns θ_1, V, and W as functions of the known quantities ρ_1, ρ, c_1, c, and θ. Note that if the two media are the same, then $V = 0$ and $W = 1$. Thus, there is no reflection, and the transmission is 100%; that is, the incident wave continues to move along the original path. As $\theta \Rightarrow \Pi/2$ (grazing incidence), then $V \Rightarrow 1$ and $W \Rightarrow 0$. This says that there is no wave transmitted to the second medium at grazing incidence of the incident wave. For θ such that $(\rho_1/\rho)\cos\theta = (k_1/k)\cos\theta$, the reflection coefficient vanishes, and there is complete transmission similar to the case in which the two media are the same.

2.6.3 Interference

If the pressure in two different acoustic waves is p_1, p_2 and the velocity of the waves is u_1, u_2, respectively, then the intensity I for each of the waves is

$$I_1 = p_1u_1 \quad I_2 = p_2u_2 \tag{2.57}$$

When the waves are combined, the pressure p in the combined wave is

$$p = p_1 + p_2 \tag{2.58}$$

and the velocity u in the combined wave is

$$u = u_1 + u_2 \tag{2.59}$$

The intensity of the combined wave is

$$\begin{aligned} I &= pu \\ &= (p_1 + p_2)(u_1 + u_2) = p_1u_1 + p_2u_2 + p_2u_1 \\ &\quad + p_1u_2 \\ &= I_1 + I_2 + (p_2u_1 + p_1u_2) \end{aligned} \tag{2.60}$$

Equation 2.60 states that the sum of the intensities of the two waves is not merely the sum of the intensities of each of the waves, but that there is an extra term. This term is called the interference term. The phenomenon that the superposition principle does not hold for intensity in linear acoustics is known as interference. If both u_1 and u_2 are positive, then what results is called "constructive interference." If $u_1 = -u_2$, then $I = 0$ and what results is called destructive interference.

2.6.4 Scattering

The discussion of reflection, refraction, and transmission was limited to waves that impinged on a flat infinite surface such as the interface between two fluids. In those cases, the phenomena of reflection, refraction, and transmission were clear-cut, and the various phenomena could be separated.

If the acoustic wave encounters a finite object, the processes are not so clear-cut and cannot be separated. The process of scattering actually involves reflection, refraction, and transmission combined, but it is called scattering because the wave scatters from the object in all directions.

Consider the classical two-dimensional problem of a plane wave impinging on a rigid cylinder, as shown in Figure 2.7. The intensity of the scattered wave can be written as

$$I_s \approx \left(\frac{2I_o a}{\pi r}\right)|\psi_s(\theta)|^2 \tag{2.61}$$

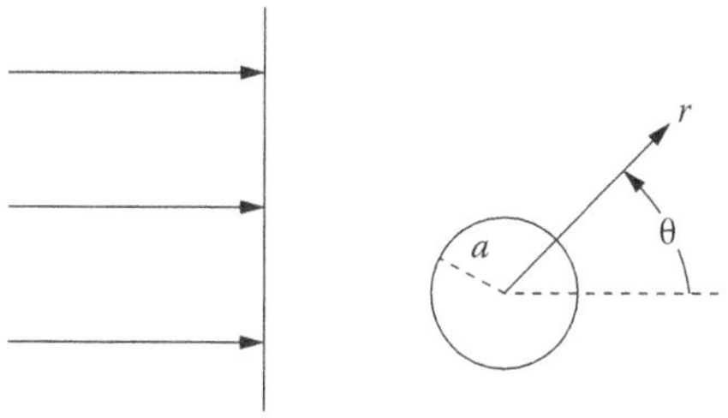

FIGURE 2.7
Plane wave impinging on a rigid cylinder. I_0 = intensity of incident plane wave; a = radius of cylinder; r, θ = cylindrical coordinates of field point.

where

I_0 is the intensity of the incident wave ($I_0 = P_0^2/2\rho_0 c_0$, where P_0 is the pressure in the incident wave, ρ_0 is the density of the medium, and c_0 is the sound velocity in the medium)

$\psi_s(\theta)$ is a distribution function

Several interesting cases can be noted. If $ka \Rightarrow 0$, then the wavelength of the sound is very large compared with the radius of the cylinder, and the scattered power goes to zero. This means that the sound travels as if the object were not present at all. If the wavelength is very small compared with the cylinder radius, it can be shown that most of the scattering is in the backward direction in the form of an echo or reflection, in the same manner as would occur at normal incidence of a plane wave on an infinite plane. Thus, for small ka (low frequency), there is mostly forward scattering, and for large ka (high frequency), there is mostly backscattering.

2.7 Free and Confined Waves

2.7.1 Propagating Waves

The acoustic wave equation states that the pressure satisfies the following equation:

$$c^2\nabla^2 p = \frac{\partial^2 p}{\partial t^2} \tag{2.62}$$

For illustrative purposes, consider the one-dimensional case in which the waves are traveling in the x direction. The equation satisfied in this case is

$$c^2\frac{\partial^2 p}{\partial x^2} = \frac{\partial^2 p}{\partial t^2} \tag{2.63}$$

The most general solution to this equation can be written in the form

$$p = f_1(x+ct) + f_2(x-ct) \tag{2.64}$$

This solution consists of two free-traveling waves moving in opposite directions.

2.7.2 Standing Waves

Consider the waves described immediately before, but limit the discussion to harmonic motion of frequency ω. One of the waves takes the form of

$$p = A\cos(kx - \omega t) \tag{2.65}$$

where $k = \omega/c = 2\Pi/\lambda$ and λ = wave length of the sound. This equation can also be written in the form of

$$p = A\cos k(x - ct) \tag{2.66}$$

If this wave hits a rigid barrier, another wave of equal magnitude is reflected back toward the source of the wave motion. The reflected wave is of the form

$$p = A\cos k(x + ct) \tag{2.67}$$

If the source continues to emit waves of the form of Equation 2.66 and reflections of the form of Equation 2.67 come back, then the resulting pattern is a superposition of the waves—that is,

$$p = A\cos(kx + \omega t) + A\cos(kx - \omega t) \tag{2.68}$$

or

$$p = 2A\ \cos kx \cos \omega t \tag{2.68a}$$

The resultant wave pattern no longer has the characteristics of a traveling wave. The pattern is stationary and is known as a standing wave.

2.7.3 Reverberation

When traveling waves are sent out in an enclosed space, they reflect from the walls and form a standing wave pattern in the space. This is a very simple description of a very complicated process in which waves impinge on the walls from various angles, reflect, and impinge again on another wall, and so on. The process of reflection takes place continually, and the sound is built up into a sound field known as a reverberant field. If the source of the sound is cut off, the field decays. The amount of time that it takes for the sound energy density to decay by a factor of, for example, 60 dB is called the reverberation time of the room. The sound energy density concept was used by Sabine in his fundamental discoveries on reverberation. He found that sound fills a reverberant room in such a way that the average energy per unit volume (i.e., the energy density) in any region is nearly the same as in any other region.

The amount of reverberation depends on how much sound is absorbed by the walls in each reflection and in the air. The study of room reverberation and the answering of questions, such as how much the sound is absorbed by people in the room and other absorbers placed in the room, are included in the field of architectural acoustics. The acoustical design of concert halls or any structures in which sound and reverberation are of importance is a specialized and intricate art.

2.7.4 Wave Guides and Ducts

When a wave is confined in a pipe or duct, the duct is known as a wave guide because it prevents the wave from moving freely into the surrounding medium. In discussing normal mode theory in connection with underwater sound propagation, the boundary conditions were stipulated on the surface and bottom. The problem thus became one of propagation in a wave guide.

One wave guide application that leads to interesting implications when coupled and uncoupled systems are considered is the propagation of axially symmetric waves in a fluid-filled elastic pipe. If axially symmetric pressure waves of magnitude P_0 and frequency ω are sent out from a plane wave source in a fluid-filled circular pipe, the pressure at any time t and at any location x from the source can be written as follows:

$$p \approx p_0\left[1 + i\omega\left(\frac{r^2}{2ac\zeta_t}\right)\right]e^{-(xk_t/a)+i[\omega/c+(\sigma_t/a)]x-i\omega t} \tag{2.69}$$

where
r is the radial distance from the center of the pipe to any point in the fluid
a is the mean radius of the pipe
c is the sound velocity in the fluid inside the pipe
ω is the radian frequency of the sound
x is the longitudinal distance from the disturbance to any point in the fluid
z_t is the impedance of the pipe such that

$$\frac{1}{z_t} = \frac{1}{\rho c \zeta_t} = \frac{1}{\rho c}(k_t - i\sigma_t) \quad (2.70)$$

where $k_t/\rho c$ is the conductance of the pipe and $\sigma_t/\rho c$ is the susceptance of the pipe. The approximate value of the wave velocity υ down the tube is

$$\upsilon \approx c[1 - \sigma_t(\lambda/2\pi a)] \quad (2.71)$$

If the tube wall were perfectly rigid, then the tube impedance would be infinite ($\sigma_t \Rightarrow 0$) and the velocity would be c. The attenuation is given by the factor $e^{ik_t x/a}$. If the tube were perfectly rigid $k_t = 0$, then the attenuation would be zero. If the tube is flexible, then energy gradually leaks out as the wave travels and the wave in the fluid attenuates. This phenomenon is used extensively in trying to reduce sound in tubes and ducts by using acoustic liners. These acoustic liners are flexible and absorb energy as the wave travels down the tube.

One critical item must be mentioned at this point. It has been assumed that the tube impedance (or conductance and susceptance) can be calculated independently. This is an assumption that can lead to gross errors in certain cases. It will be discussed further when coupled systems are considered.

The equation for an axisymmetric wave propagating in a rigid circular duct or pipe is as follows:

$$p(r,z) = p_{mo} J_0(\alpha_{om} r/a) e^{i(\gamma_{om} z - \omega t)} \quad (2.71a)$$

where
p_{mo} is the amplitude of the pressure wave
r is the radial distance from the center of the pipe to any point
a is the radius of the pipe
z is the distance along the pipe
ω is the radian frequency
t is time

γ_{om} and α_{om} are related by the following formula:

$$\gamma_{om} = [k^2 - (\alpha_{om}/a)^2]^{1/2} \quad (2.71b)$$

J_0 is the Bessel function of order 0.

$$k = 2\pi f/c_i \quad (2.71c)$$

In the preceding relation, f is the frequency of the wave and c_i is the sound velocity in the fluid inside the pipe (for water, this sound velocity is about

5000 ft/sec and for air it is about 1100 ft/sec). The values of α_{om} for the first few m are the following:

$$m = 0 \quad \alpha_{00} = 0$$
$$m = 1 \quad \alpha_{01} = 3.83 \tag{2.71d}$$
$$m = 2 \quad \alpha_{02} = 7.02$$

If $k < \alpha_{om}/a$, then γ_{om} is a pure imaginary number and the pressure takes the form of

$$p(r, z) = P_{mo} J_0(\alpha_{om} r/a) e^{-\gamma_{om} z} e^{-i\omega t} \tag{2.71e}$$

which is the equation for a decaying wave in the z direction. For frequencies that give $k < \alpha_{om}/a$, no wave is propagated down the tube. Propagation takes place only for frequencies in which $k > \alpha_{om}/a$. Since $\alpha_{oo} = 0$, propagation always takes place for this mode. The frequency at which γ_{om} is 0 is called the cutoff frequency and is as follows:

$$f_{om} = c_i \alpha_{om}/2\pi a \tag{2.71f}$$

For frequencies below the cutoff frequency, no propagation takes place. For general asymmetric waves, the pressure takes the form of

$$p(r, z, \phi) = P_{nm} J_n(\alpha_{nm} r/a) e^{i(\gamma_{nm} z - \omega t)} \cos n\phi$$
$$\text{where } \gamma_{nm} = [k^2 - (\alpha_{nm}/a)^2]^{1/2}, f_{nm} = c_i \alpha_{nm}/2\pi a \tag{2.71g}$$

A few values for α_{nm} for $n > 0$ are as follows:

$$\alpha_{10} = 1.84 \quad \alpha_{11} = 5.31 \quad \alpha_{12} = 8.53$$
$$\alpha_{20} = 3.05 \quad \alpha_{21} = 6.71 \quad \alpha_{22} = 9.97$$

It is seen that only $\alpha_{oo} = 0$; this is the only mode that propagates at all frequencies, regardless of the size of the duct.

Consider a 10-in. diameter pipe containing water. The lowest cutoff frequency greater than the 00 mode is 3516 Hz. Thus, nothing will propagate below 3516 Hz except the lowest axisymmetric mode (i.e., the 00 mode). If the pipe were 2 in. in diameter, then nothing would propagate in the pipe below 17,580 Hz except the 00 mode. This means that in a great many practical cases, no matter what is exciting the sound inside the duct, only the lowest axisymmetric mode (00 mode) will propagate in the duct.

2.8 Sound Radiation and Vibration

2.8.1 Helmholtz Integral, Sommerfeld Radiation, and Green's Function

In this section, acoustic fields that satisfy the wave equation of linear acoustics are of interest:

$$c^2\nabla^2 p = \partial^2 p/\partial t^2 \tag{2.72}$$

where p is the pressure in the field at point P and at time t. For sinusoidal time-varying fields,

$$p(P,t) = p(P)e^{i\omega t} \tag{2.73}$$

so that p satisfies the Helmholtz equation:

$$\nabla^2 p + k^2 p = 0 \quad k^2 = \omega^2/c^2 \tag{2.74}$$

This Helmholtz equation can be solved in general form by introducing an auxiliary function called Green's function. First, let φ and ψ be any two functions of space variables that have first and second derivatives on S and outside S (see Figure 2.8). Let V_0 be the volume between S_0 and the boundary at ∞. Green's theorem states that within the volume V_0,

$$\int_S \left(\varphi\frac{\partial\psi}{\partial n} - \psi\frac{\partial\varphi}{\partial n}\right) dS = \int_{V_0} (\varphi\nabla^2\psi - \psi\nabla^2\varphi)\, dV_0 \tag{2.75}$$

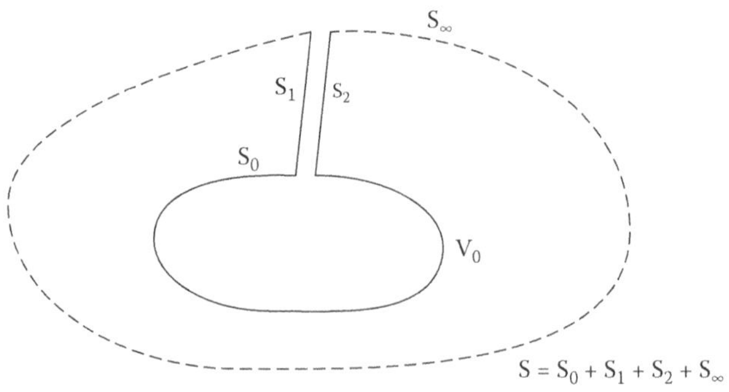

FIGURE 2.8
The volume and boundary surface. S_∞ = surface at ∞; S_0 = radiating surface; V_0 = volume between S_∞ and S_0; S_1, S_2 = surfaces connecting S_0 to S_∞.

where S denotes the entire boundary surface and V_0 the entire volume of the region outside S_0.

In Equation 2.75, $\partial/\partial n$ denotes the normal derivative at the boundary surface. Rearrange the terms in Equation 2.75 and subtract $\int V_0 k^2 \varphi\psi dV_0$ from each side; the result is

$$\int_S \varphi \frac{\partial \psi}{\partial n} dS - \int_{V_0} \varphi(\nabla^2\psi + k^2\psi) dV_0 = \int_S \psi \frac{\partial \varphi}{\partial n} dS - \int_{V_0} \psi(\nabla^2\varphi + k^2\varphi) dV_0 \tag{2.76}$$

Now, choose φ as the pressure p in the region V_0; thus,

$$\nabla^2 p + k^2 p = \nabla^2\varphi + k^2\varphi = 0 \tag{2.77}$$

Choose ψ as a function that satisfies

$$\nabla^2\psi + k^2\psi = \delta(P - P') \tag{2.78}$$

where $\delta(P - P')$ is a δ function of field points P and P'. Choose another symbol for ψ—that is,

$$\psi = g(P, P', \omega) \tag{2.79}$$

By virtue of the definition of the δ function, the following is obtained:

$$\int_{V_0} \varphi(P')\delta(P - P')dP' = \varphi(P) \tag{2.80}$$

Thus, Equation 2.76 becomes

$$\int_S p(S,\omega)\frac{\partial g(P,S,\omega)}{\partial n} - p(P,\omega) = \int_S g(P,S,\omega)\frac{\partial p(S,\omega)}{\partial n} dS \tag{2.81}$$

or

$$p(P,\omega) = \int_S \left[p(S,\omega)\frac{\partial g(P,S,\omega)}{\partial n} - g(P,S,\omega)\frac{\partial p(S,\omega)}{\partial n} \right] dS \tag{2.82}$$

It is now clear that the arbitrary function ψ was chosen so that the volume integral would reduce to the pressure at P. The function g is the Green's function, which thus far is a completely arbitrary solution of

$$\nabla^2 g(P,P',\omega)+k^2 g(P,P',\omega)=\delta(P-P') \tag{2.83}$$

For this Green's function to be a possible solution, it must satisfy the condition that there are only outgoing traveling waves from the surface S_o to ∞ and that no waves are coming in from ∞. Sommerfield formulated this condition as follows:

$$\lim_{r\to\infty} r\left(\frac{\partial g}{\partial r}-ikg\right)=0 \tag{2.84}$$

where r is the distance from any point on the surface to any point in the field. A solution that satisfies Equations 2.83 and 2.84 can be written as follows:

$$g=\frac{1}{4\pi}\frac{e^{ikr}}{r} \tag{2.85}$$

This function is known as the free space Green's function. Thus, Equation 2.82 can be written in terms of this free space Green's function as follows:

$$\begin{aligned} p(P,\omega)=\frac{1}{4\pi}\int_S\Bigg[&p(S,\omega)\frac{\partial}{\partial n}\left(\frac{e^{ikr}}{r}\right)\\ &-\frac{e^{ikr}}{r}\frac{\partial p(S,\omega)}{\partial n}\Bigg]dS \end{aligned} \tag{2.86}$$

Several useful alternative forms of Equations 2.82 and 2.86 can be derived. If a Green's function can be found whose normal derivative vanishes on the boundary S_0, then

$$\frac{\partial g}{\partial n}(P,S,\omega)=0 \quad \text{on } S_0$$

Equation 2.82 then becomes

$$p(P,\omega)=-\int_S g(P,S,\omega)\frac{\partial p}{\partial n}dS \tag{2.87}$$

Alternatively, if a Green's function can be found that itself vanishes on the boundary S_0, then $g(P, S, \omega) = 0$ on S_0 and Equation 2.82 becomes

$$p(P,\omega) = \int_S p(S,\omega)\frac{\partial g(P,S,\omega)}{\partial n}dS \tag{2.88}$$

From Newton's law,

$$\partial p/\partial n = -\rho\ddot{w}_n \tag{2.89}$$

where $\ddot{w}_n$ is the normal acceleration of the surface. Thus, Equation 2.87 can be written as

$$p(P,\omega) = \rho\int_S g(P,S,\omega)\ddot{w}_n\, dS \tag{2.90}$$

If γ is the angle between the outward normal to the surface at S and the line drawn from S to P, then (assuming harmonic motion) Equation 2.86 can be written in the form of

$$p(P,\omega) = \frac{1}{4\pi}\int_S (\rho\ddot{w}_n(s) - ikp(s)\cos\gamma)\frac{e^{ikr}}{r}dS \tag{2.91}$$

Since $p(s)$ is unknown, Equation 2.86 or, alternatively, Equation 2.91 is an integral equation for p.

An interesting limiting case can be studied. Assume a low-frequency oscillation of the body enclosed by S_0. For this case, k is very small and Equation 2.91 reduces to

$$p(P,\omega) = \frac{1}{4\pi r}\int_S \rho\ddot{w}_n\, ds \tag{2.92}$$

or

$$p(P,\omega) = \rho\ddot{V}/4\pi R \tag{2.93}$$

where $\ddot{V}$ is the volume acceleration of the body. For some cases of slender bodies vibrating at low frequency, it can be argued that the term involving p in Equation 2.91 is small compared with the acceleration term. For these cases,

$$p(P,\omega) \approx \frac{1}{4\pi}\int_S \rho\ddot{w}_n(S)\frac{e^{ikr}}{r}ds \tag{2.94}$$

2.8.2 Rayleigh's Formula for Planar Sources

It was stated in the last section that if a Green's function could be found whose normal derivative vanishes at the boundary, then the sound pressure radiated from the surface could be written as Equation 2.87 or, alternatively, using Equation 2.89,

$$p(P,\omega)=\rho\int_S g(P,S,\omega)\ddot{w}_n(S)ds \tag{2.95}$$

It can be shown that such a Green's function for an infinite plane is exactly twice the free space Green's function—that is,

$$g(P,S,\omega)=2\times\frac{1}{4\pi}\frac{e^{ikr}}{r} \tag{2.96}$$

The argument here is that the pressure field generated by two identical point sources in free space displays a zero derivative in the direction normal to the plane of symmetry of the two sources. Substituting Equation 2.96 into Equation 2.95 gives

$$p=\frac{p}{2\pi}\int_S\frac{e^{ikr}}{r}\ddot{w}_n(S)ds \tag{2.97}$$

Equation 2.97 is known as Rayleigh's formula for planar sources.

2.8.3 Vibrating Structures and Radiation

To make it clear how the preceding relations can be applied, consider the case of a slender vibrating cylindrical surface at low frequency such that Equation 2.94 applies. The geometry of the problem is shown in Figure 2.9,

$$\begin{aligned}\frac{e^{ikr}}{r}&=\frac{e^{ikR_1}}{R_1}e^{-ik(\mathbf{a}_{R_1}\cdot\mathbf{R}_{S_1})}\\ \mathbf{a}_{R_1}\cdot\mathbf{R}_{S_1}&=z_o\cos\theta_1+x_o\sin\theta_1\cos\varphi_1\\ &\quad+y_o\sin\theta_1\sin\varphi_1\end{aligned} \tag{2.98}$$

where x_o, y_o, and z_o are the rectangular coordinates of a point on the vibrating surface of the structure; $\mathbf{R}s_1$ is the radius vector to point S_1 on the surface;

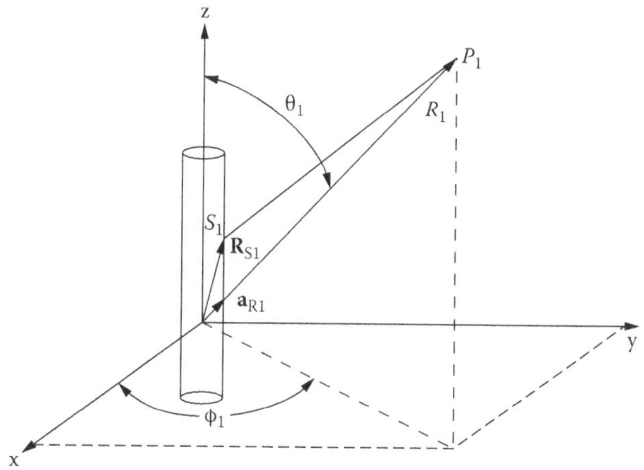

FIGURE 2.9
Geometry of the vibrating cylinder. $\mathbf{R}s_1$ = radius vector to point S_1 on the cylinder surface; R_1 = radius vector to the far field point; $\mathbf{a}_{R1}$ = unit vector in the direction of R_1; P_1 = far field point (with spherical coordinates R_1, φ_1, θ_1). (Reprinted with permission from Greenspon, J. E. 1967. *Journal of Acoustical Society of America* 41 (5): 1203.)

R_1, φ_1, and θ_1 are the spherical coordinates of point P_1 in the far field; and $\mathbf{a}_{R1}$ is a unit vector in the direction of R_1 (the radius vector from the origin to the far field point). Thus, $\mathbf{a}_{R1}\,\mathbf{R}_{S1}$ is the projection of $\mathbf{R}s_1$ on $R1$, making $R_1 - \mathbf{a}_{R1}\,\mathbf{R}s_1$ the distance from the far field point to the surface point.

Assume the acceleration distribution of the cylindrical surface to be that of a freely supported shell subdivided into its modes—that is,

$$\ddot{w}_n(S) = \sum_{m=1}^{\infty}\sum_{q=0}^{\infty}(A_{mq}\cos q\varphi_1 + B_{mq}\sin q\varphi)\sin\frac{m\pi z}{l} \tag{2.99}$$

Expression 2.99 is a half-range Fourier expansion in the longitudinal z-direction (which is chosen as the distance along the generator) between bulkheads, which are distance 1 apart. Expression 2.99 is a full-range Fourier expansion in the peripheral ~p direction. It is known that such an expression does approximately satisfy the differential equations of shell vibration and practical end conditions at the edges of the compartment. Substitution of Equations 2.98 and 2.99 into Equation 2.94 and integrating results in

$$p(P_1,\omega) = \frac{\rho_o e^{ikR_1}}{4\pi R_1}\sum_{m=1}^{\infty}\sum_{q=0}^{\infty}$$

$$\{A_{mq}\cos q\varphi_1 + B_{mq}\sin q\varphi_1\}$$

$$\times J_q(ka\sin\theta_1)2\pi ali^q$$

$$\times\left\{\left[\frac{1}{2(m\pi - kl\cos\theta_1)}\right.\right.$$

$$\left.+\frac{1}{2(m\pi + kl\cos\theta_1)}\right]$$

$$-\left[\frac{\cos(m\pi - kl\cos\theta_1)}{2(m\pi - kl\cos\theta_1)}\right. \tag{2.99a}$$

$$\left.+\frac{\cos(m\pi + kl\cos\theta_1)}{2(m\pi + kl\cos\theta_1)}\right]$$

$$-i\left[\frac{\sin(m\pi - kl\cos\vartheta_1)}{2(m\pi - kl\cos\theta_1)}\right.$$

$$\left.\left.-\frac{\sin(m\pi + kl\cos\vartheta_1)}{2(m\pi + kl\cos\theta_1)}\right]\right\}$$

Consider the directivity pattern in the horizontal plane of the cylindrical structure—that is, at $\varphi_1 = \Pi/2$—and let us examine the A_{mq} term in the preceding series. After some algebraic manipulation, the amplitude of the pressure can be written as follows:

For $m\pi \neq kl\cos\theta_1$,

$$I_{mq} = \frac{p_{mq}/2\pi al}{\dfrac{A_{mq}\rho_o e^{ikR_1}}{4\pi R_1}\cos q\theta_1}$$

$$= \frac{\sqrt{2}m\pi J_q(ka\sin\theta_1)\sqrt{1-(-1)^m\cos(kl\cos\theta_1)}}{(m\pi)^2 - (kl\cos\theta_1)^2}$$

For $m\pi = kl\cos\theta_1$,

$$I_{mq} = \tfrac{1}{2}J_q(ka\ \sin\theta_1) \tag{2.100}$$

where J_q is the Bessel function of order q.

Figure 2.10 shows the patterns of the far field pressure for various values of *ka*, *m*, and *q*. A source pattern is defined as one that is uniform in all directions. A dipole pattern has two lobes, a quadrupole pattern has four lobes, and so on. Note that for $ka = 0.1$, $q = 1$, and $m = 1, 3, 5$, all show dipole-type radiation. In general, Figure 2.10 shows how the multipole contributions depend upon the spatial pattern of the acceleration of the structure.

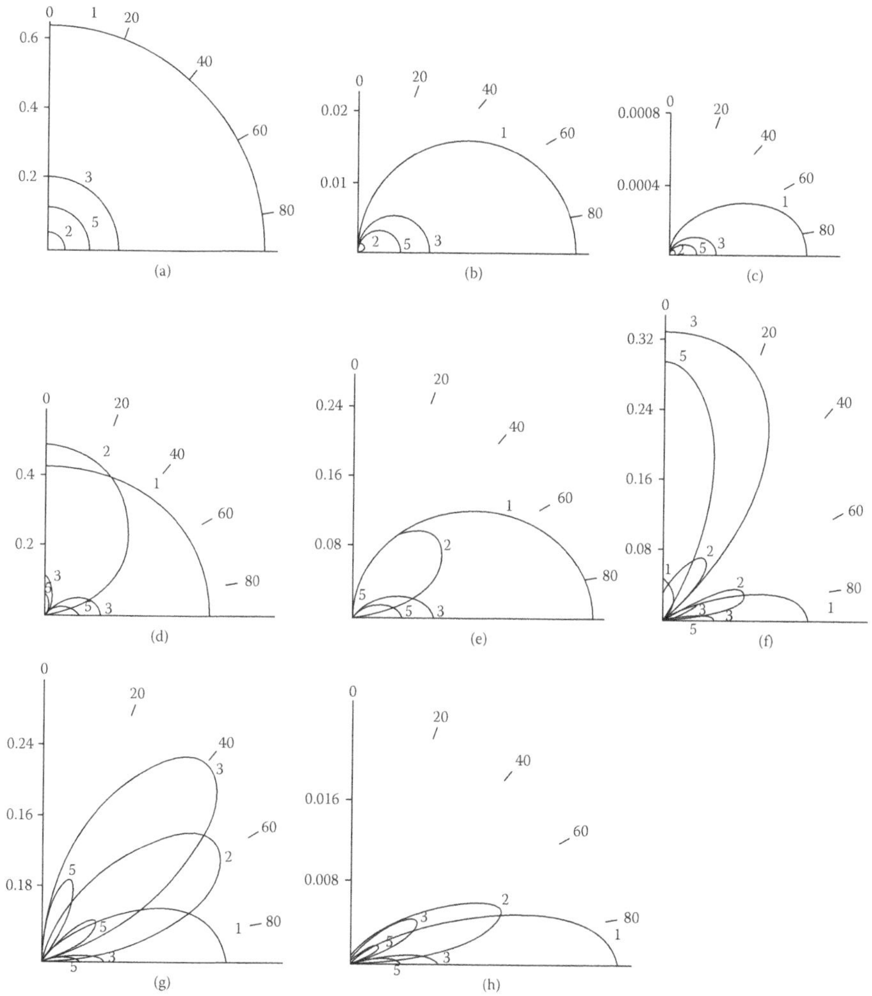

FIGURE 2.10
Horizontal directivity patterns for a cylinder in which $L/a = 4$, where L is length of cylinder and a is radius of cylinder. The plots show I_{mq} as a function of θ_1 at $\varphi_1 = \pi/2$ for various values of *ka*, *m*, and *q*. (a) $ka = 0.1$, $q = 0$; (b) $ka = 0.1$, $q = 1$; (c) $ka = 0.1$, $q = 2$; (d) $ka = 1.0$, $q = 0$; (e) $ka = 1.0$, $q = 1$; (f) $ka = 3.0$, $q = 0$; (g) $ka = 3.0$, $q = 1$; (h) $ka = 3.0$, $q = 5$. The numbers shown on the figure are the values of *m*. (Reprinted with permission from Greenspon, J. E. 1967. *Journal of Acoustical Society of America* 41 (5): 1205.)

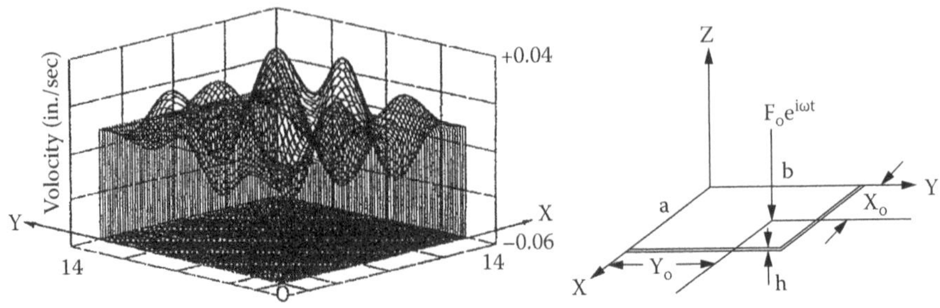

FIGURE 2.11
Real part of the velocity of the plate.

For low frequencies where $kl = m\pi$, it is seen that (noting that $\pi l/\lambda = kl/2$) for m even,

$$
\begin{aligned}
I_{mq} &\approx \frac{2J_q(ka\sin\theta_1)\sin\left(\dfrac{\pi l}{\lambda}\cos\theta_1\right)}{m\pi} \\
\text{for } m \text{ odd,}\quad I_{mq} &\approx \frac{2J_q(ka\sin\theta_1)\cos\left(\dfrac{\pi l}{\lambda}\cos\theta_1\right)}{m\pi}
\end{aligned}
\tag{2.101}
$$

Thus, at low frequencies (i.e., for $kl \ll m\pi$), the structural modes radiate as though there were two sources at the edges of the compartment (i.e., one apart). What results is the directivity pattern for two point sources at the edges of the compartment modified by $Jq(ka\ \sin\theta_1)$. If the longitudinal mode number m is even, then the sources are 180° out of phase, and if m is odd, the sources are in phase. Such modes are called edge modes.

2.8.4 Vibration of Flat Plates in Water

Consider a simply supported flat rectangular elastic plate placed in an infinite plane baffle and excited by a point force, as shown in Figure 2.11. The plate is made of aluminum and is square with each side being 13.8 in. long. Its thickness is 1/4 in., and it has a damping loss factor of 0.05. The force is equal to 1 lb, has a frequency of 3000 cps, and is located at $x_o = 9$ in., $y_o = 7$ in. If the plate is stopped at the instant when the deflection is a maximum, the velocity pattern would be as shown in Figure 2.11. Since the velocity is just the frequency multiplied by the deflection, this is an instantaneous picture of the plate.

2.9 Coupling of Structure/Medium (Interactions)

2.9.1 Coupled versus Uncoupled Systems

When a problem involving the interaction between two media can be solved without involving the solution in both media simultaneously, the problem is said to be uncoupled. One example of a coupled system is a vibrating structure submerged in a fluid. Usually, the amplitude of vibration depends on the dynamic fluid pressure, and the dynamic pressure in the fluid depends on the amplitude of vibration. In certain limiting cases, the pressure on the structure can be written as an explicit function of the velocity of the structure. In these cases, the system is said to be uncoupled.

Another example is a pipe containing an acoustic liner. Sometimes it is possible to represent the effect of the liner by an acoustic impedance, as described in a previous section of this chapter. Such a theory was offered by Morse.[14] Implicit in Morse's theory is the assumption that the motion of the surface between the liner and the fluid depends only on the acoustic impedance and the local acoustic pressure, rather than on the acoustic pressure elsewhere. This is associated with the concept of "local" and "extended" reactions. In a truly coupled system, the motion of the surface depends on the distribution of acoustic pressure and, conversely, the acoustic pressure depends on the distribution of motion.

Thus, the reaction of the surface and the pressure produced are interrelated at all points. The motion of the surface at point A is not governed only by the pressure at point A. There is motion at A due to pressure at B and, conversely, motion at B due to pressure at A. The alternative is to solve the completely coupled problem of the liner and fluid.

In aeroelastic or hydroelastic problems, it is necessary to solve the coupled problem of structure and fluid because the stability of the system usually depends on the feeding of energy from the fluid to the structure. Similarly, in acoustoelastic problems such as soft duct liners in pipes, the attenuation of the acoustic wave in the pipe is dependent upon the coupling of this acoustic wave with the wave in the liner. Similar problems have been encountered with viscoelastic liners in water-filled pipes.

References

1. Ahluwalia, D. S., and Keller, J. B. 1977. Exact and asymptotic representations of the sound field in a stratified ocean. *Lecture Notes in Physics* 70:14.
2. Brekhovskikh, L., and Lysanov, Y. 1982. *Fundamentals of ocean acoustics*. Berlin: Springer–Verlag.

3. Camp, L. 1970. *Underwater acoustics.* New York: Wiley (Interscience).
4. Dowling, A. P., and Ffowcs Williams, J. E. 1983. *Sound and sources of sound.* New York: Halsted Press.
5. Geers, T. L. 1978. Doubly asymptotic approximations for transient motions of submerged structures. *Journal of Acoustical Society of America* 64 (5): 1500.
6. Greenspon, J. E. 1981. In *Encyclopedia of physics,* ed. R. G. Lerner and G. L. Trigg, 14. Reading, MA: Addison–Wesley.
7. Greenspon, J. E. 1983. The theory of cylindrical acoustic liners with longitudinal holes based on the three dimensional elasticity theory. Report NS-83-1, contract N61533-82-M-2871. JG Engineering Research Associates, Baltimore, MD.
8. Huang, H. 1981. Approximate fluid–structure interaction theories for acoustic echo signal prediction. NRL memo. rep. 4661. Naval Research Laboratory, Washington, DC.
9. Junger, M. C., and Felt, D. 1972. *Sound structures and their interaction.* Cambridge, MA: MIT Press.
10. Liszka, L. 1978. Long-distance focusing of Concorde sonic boom. *Journal of Acoustical Society of America* 64 (2): 631–635.
11. Pierce, A. D. 1981. *Acoustics: An introduction to its physical principles and applications.* New York: McGraw–Hill.
12. Pierce, A. D. 1974. Diffraction of sound around corners and over wide barriers. *Journal of Acoustical Society of America* 55 (5): 941–955.
13. Greenspon, J. E. 1992. Acoustics, linear. In *Encyclopedia of physical science and technology,* vol. 1, 114–152. New York: Academic Press, Inc.–Elsevier.
14. Morse, P. M. 1948. *Vibration and sound.* New York: McGraw–Hill Book Co., Inc.

Glossary

Acoustics: The science of sound, its generation, transmission, reception, and effects. Linear acoustics is the study of the physical phenomena of sound in which the ratio of density change to static density is small, typically much less than 0.1. A sound wave is a disturbance that produces vibrations of the medium in which it propagates.

Attenuation: Reduction in amplitude of a wave as it travels.

Condensation: Ratio of density change to static density.

Coupling: Mutual interaction between two wave fields.

Diffraction: Bending of waves around corners and over barriers.

Dispersion: Dependence of velocity on frequency, manifested by distortion in the shape of a disturbance.

Elastic waves: Traveling disturbances in solid materials.

Ergodic: Statistical process in which each record is statistically equivalent to every other record. Ensemble averages over a large number of records at fixed times can be replaced by corresponding time averages on a single representative record.

Impedance: Pressure per unit velocity.
Interaction: Effect of two media on each other.
Medium: Material through which a wave propagates.
Nondispersive medium: Medium in which the velocity is independent of frequency and the shape of the disturbance remains undistorted.
Normal mode: Shape function of a wave pattern in transmission.
Propagation: Motion of a disturbance characteristic of radiation or other phenomena governed by wave equations.
Ray: Line drawn along the path along which the sound travels, perpendicular to the wave front.
Reflection: Process of a disturbance bouncing off an obstacle.
Refraction: Change in propagation direction of a wave with change in medium density.
Reverberation: Wave pattern set up in an enclosed space.
Scattering: Property of waves in which a sound pattern is formed around an obstacle enveloped by an incoming wave.
Sommerfield radiation condition: Equation stating that waves must go out from their source toward infinity and not come in from infinity.
Standing waves: Stationary wave pattern.
Wave guide: Structure or channel along which the wave is confined.

3

Approximations for Added Mass and Radiation Damping

One of the most difficult tasks facing the author in writing this book was to select a simple but reasonably accurate approximation for handling fluid loading on the structure. The purpose of this book is to give the reader complete Green's functions, including a computer program. The author's intention is to provide general types of solutions with varying boundary conditions. Thus, the following approximations have been used. If the reader desires more accurate approximations for the fluid loading, many papers contain more accurate approximations; however, in all these cases, the computations and computer work are much more involved than the presentations given here, and the results only apply to specific systems. Some of these references are given at the end of this chapter.[1–12]

For the beam (uniform or variable) and the plate (unstiffened or stiffened), it will be assumed that the fluid loading can be represented as follows:

Fluid pressure = velocity × (impedance + structural damping factor)

where the impedance is that of a rectangular piston that has the same area as that in contact with the fluid.

As pointed out by Junger,[11] at "high" frequencies, the impedance value is resistive and equals ρc, where ρ is the fluid density and c is the sound velocity in the fluid. Figure 3.1 shows the radiation efficiency ($R/\rho c A$) for a beam as measured by Lyon and Maidanik[6] and the values calculated by using the piston assumption. The values for the nondimensional radiation efficiency ($R/\rho c A$) and the virtual mass factor ($\chi/\rho c A$) are given in Figure 3.2.[13] Figure 3.3 gives the value of the radiation factor (or radiation efficiency) as shown in Maidanik's paper,[7] with the values calculated by the piston approximation.

The virtual mass function for a simply supported rectangular plate is shown in Figure 3.4.[14] The values computed by the piston approximation are shown on this figure. The approximation is suitable for plates having aspect ratios close to unity and at higher frequencies.

For unbaffled as compared to baffled structures, if the piston approximation is used, Crane[15] shows that for circular pistons the radiation factor for an unbaffled piston is about 60% of its value for a baffled piston. The virtual mass factor for an unbaffled piston is about 80% of the value for a baffled piston. The piston approximation allows one to compute the Green's functions for many practical cases, as will be shown later in this book.

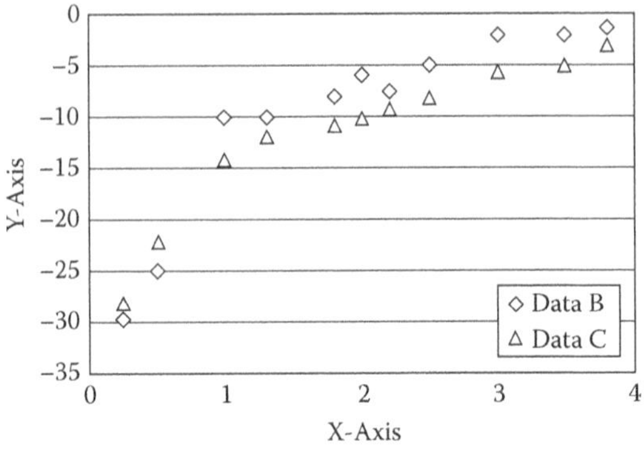

FIGURE 3.1
Radiation efficiency of a beam in a baffle. X-axis = k/k_b; Y-axis = $10\log(R/A\rho c)$; $k = \omega/c$; $k_b = n\pi/L$; L = length of beam; R = radiation resistance; A = area over which beam vibrates against fluid; ρ = mass density of fluid; c = sound velocity in fluid; Data B = Lyon–Maidanik results[6]; Data C = results of piston approximation.

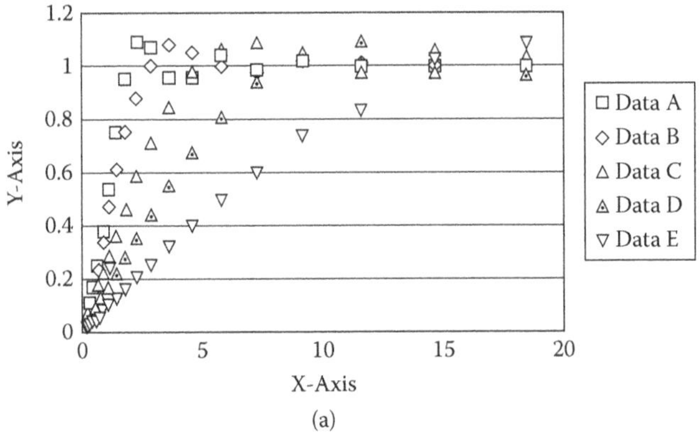

FIGURE 3.2
Resistance and reactance of a rectangular piston. Y-axis = $\bar{R}/\rho cA$ for Figures 3.2a, 3.2b = θ; Y-axis = $\bar{\chi}/\rho cA$ for Figures 3.2c, 3.2d = χ; X-axis = $k\sqrt{A/\pi}$; $k = \omega/c$; A = area; AR = aspect ratio; Data A = AR = 1:1; Data B = AR = 3:1; Data C = AR = 10:1; Data D = AR = 30:1; Data E = AR = 100:1.

For built-up structures modeled with rectangular nets (Chapter 4), the mutual impedance as well as the self-impedance used before can be used to obtain the effect of adjacent elements and therefore is a better approximation than using the self-impedance alone. The mutual impedance for rectangular pistons is given in Crane[15] and Arase.[16]

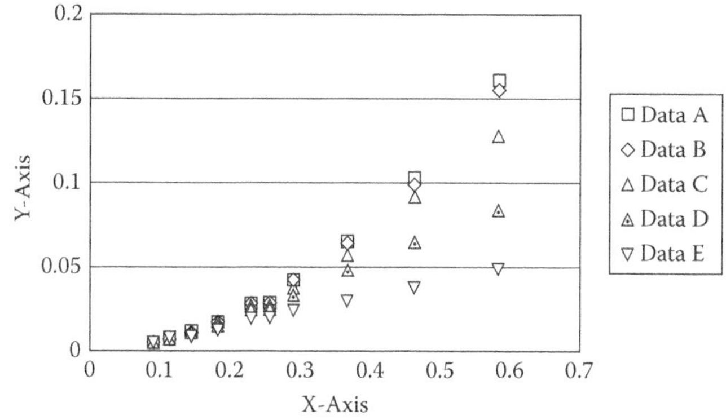

(b)

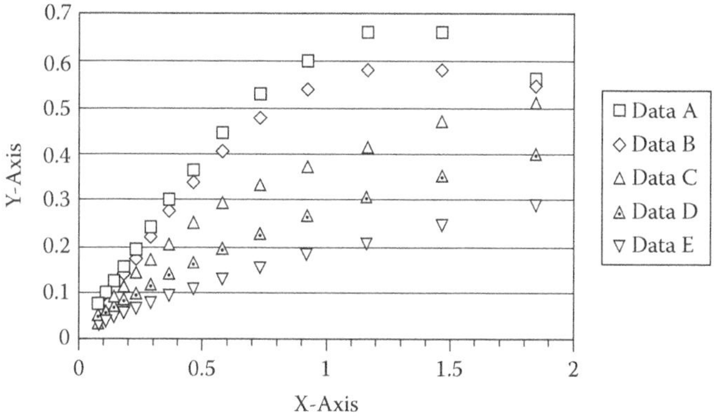

(c)

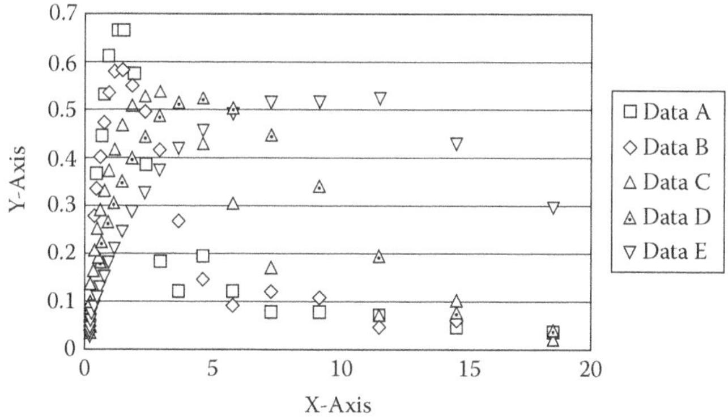

(d)

FIGURE 3.2 (Continued)
Resistance and reactance of a rectangular piston. Y-axis = $\bar{R}/\rho cA$ for Figures 3.2a, 3.2b = θ; Y-axis = $\bar{\chi}/\rho cA$ for Figures 3.2c, 3.2d = χ; X-axis = $k\sqrt{A/\pi}$; $k = \omega/c$; A = area; AR = aspect ratio; Data A = AR = 1:1; Data B = AR = 3:1; Data C = AR = 10:1; Data D = AR = 30:1; Data E = AR = 100:1.

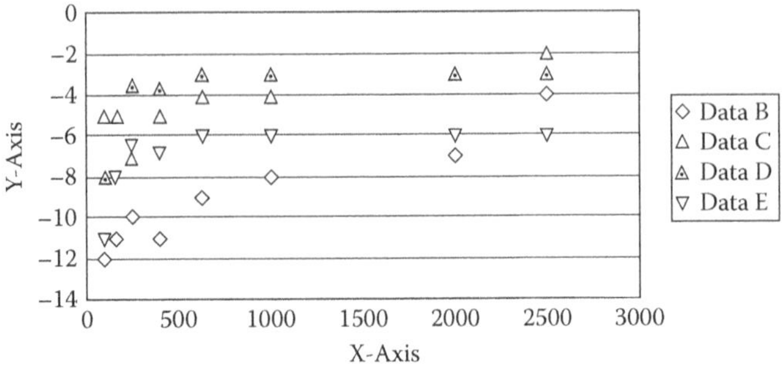

FIGURE 3.3
Radiation efficiency of a plate. X-axis = third octave center frequency; Y-axis = 10 $\log_{10}(R/2\rho cA)$; Data B = unbaffled ribbed plate[7]; Data C = baffled ribbed plate[7]; Data D = baffled piston approximation; Data E = unbaffled piston approximation.

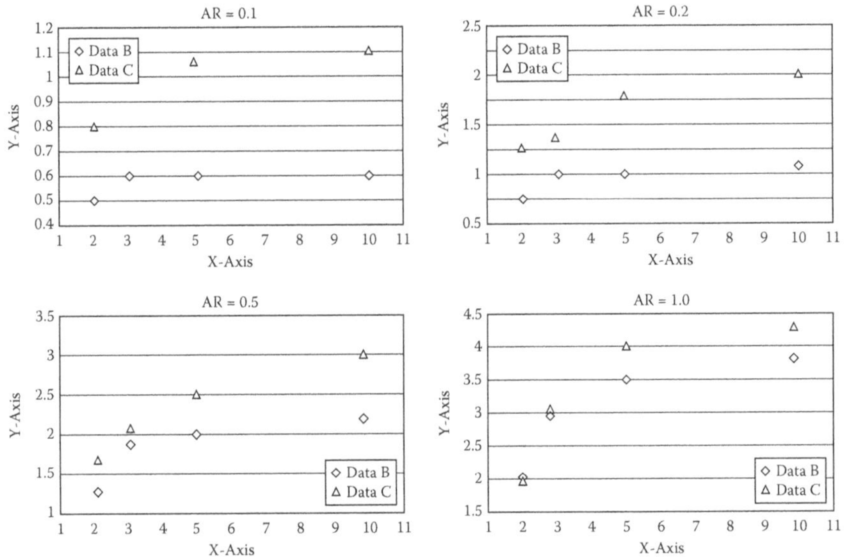

FIGURE 3.4
Virtual mass of a simply supported plate. Y-axis = virtual mass/plate mass; X-axis = acoustic wavelength/plate length; AR = aspect ratio; Data B = values from Pretlove[14]; Data C = piston approximation.

Definitions

Acoustic impedance: The ratio between acoustic pressure and fluid velocity.

Acoustic wave length: Ratio of the sound velocity in the medium to the frequency in cycles per unit time.

Radiation efficiency: The radiation resistance divided by ρcA where ρ = density of medium, c = sound velocity in medium, and A = area of the structure being studied.

Reactance: The imaginary part of the acoustic impedance; associated with the added mass of the fluid due to the motion between solid and fluid.

Resistance: The real part of the acoustic impedance; associated with the dissipation losses occurring when there is viscous movement of the fluid.

Virtual mass: The added mass due to the presence of the vibrating fluid.

References

1. Berry, A., Guyader, J. L., and Nicolas, J. 1990. A general formulation for the sound radiation from rectangular baffled plates with arbitrary boundary conditions. *Journal of Acoustical Society of America* 88 (6): 2792.
2. Berry, A. 1994. A new formulation for the vibrations and sound radiation of fluid loaded plates with elastic boundary conditions. *Journal of Acoustical Society of America* 96 (2): 889.
3. Wallace, C. E. 1972. Radiation resistance of a rectangular panel. *Journal of Acoustical Society of America* 51 (3): 946.
4. Wallace, C. E. 1972. Radiation resistance of a baffled beam. *Journal of Acoustical Society of America* 51(3): 936.
5. Mangiarotty, A. 1963. Acoustic radiation damping of vibrating structures. *Journal of Acoustical Society of America* 35 (3): 369.
6. Lyon, H., and Maidanik, G. 1962. Power flow between linearly coupled oscillators. *Journal of Acoustical Society of America* 34 (5): 623.
7. Madanik, G. 1962. Response of ribbed panels to reverberant acoustic fields. *Journal of Acoustical Society of America* 34 (6): 809.
8. Nelisse, H., Beslin, O., and Nicolas, J. 1998. A generalized approach for the acoustic radiation from a baffled or unbaffled plate with arbitrary boundary conditions immersed in a light or heavy fluid. *Journal of Sound and Vibration* 211 (2): 207.
9. Sandman, E. 1975. Motion of a three layered elastic-viscoelastic plate under fluid loading. *Journal of Acoustical Society of America* 57 (5): 1097.
10. Shuyu, L. 2002. Study on the radiation acoustic field of rectangular radiators in flexural vibration. *Journal of Sound and Vibration* 254 (3): 469.
11. Junger, C. 1987. Approaches to acoustic fluid-elastic structure interactions. *Journal of Acoustical Society of America* 82 (4): 1115.
12. Nelisse, H., Beslin, O., and Nicolas, J. 1996. Fluid-structure coupling for an unbaffled elastic panel immersed in a diffuse field. *Journal of Sound and Vibration* 198 (4): 485.

13. Bank, G., and Wright, J. R. 1990. Radiation impedance calculations for a rectangular piston. *Journal of Audio Engineering Society* 38 (5): 350.
14. Pretlove, J. 1965. Note on the virtual mass for a panel in an infinite baffle. *Journal of Acoustical Society of America* 38:266.
15. Crane, H. G. 1967. Method for the calculation of the acoustic radiation impedance of unbaffled and partially baffled piston sources. *Journal of Sound and Vibration* 5 (2): 257.
16. Arase, M. 1964. Mutual radiation impedance of square and rectangular pistons in a rigid infinite baffle. *Journal of Acoustical Society of America* 36 (8): 1521.

4

Fundamentals of Structures and Analysis of Beams

4.1 Fundamentals of Structures

4.1.1 Introduction

For the purposes of this book, structures can be divided into four groups. The first is the simplest structure, which is the one-dimensional beam discussed in this chapter. The second is the two-dimensional plate, the details of which are discussed in Chapter 5. The third is the three-dimensional shell, the cylindrical version of which is studied thoroughly in Chapter 7. Chapter 6 considers built-up shells, which are constructed of a series of plates; many complicated shell problems can be solved by using this procedure. The fourth group is the three-dimensional body, which can have different properties as a function of space and can be elastic, viscoelastic, and fluid. This is discussed in Chapter 8. In this section of Chapter 4, the basic equations for each of these cases will be discussed and it will be left to the reader to study the details in the chapters that discuss them in detail.

4.2 Beams

4.2.1 Longitudinal Vibration

A typical beam that has its cross-sectional area changing with x (the longitudinal space dimension) is shown in Figure 4.1.

Let
$u(x)$ = longitudinal deflection (Figure 4.2)
$S(x,t)$ = internal longitudinal force acting on the beam
$A(x)$ = cross-sectional area of beam at point x
$p(x,t)$ = external longitudinal load/unit length

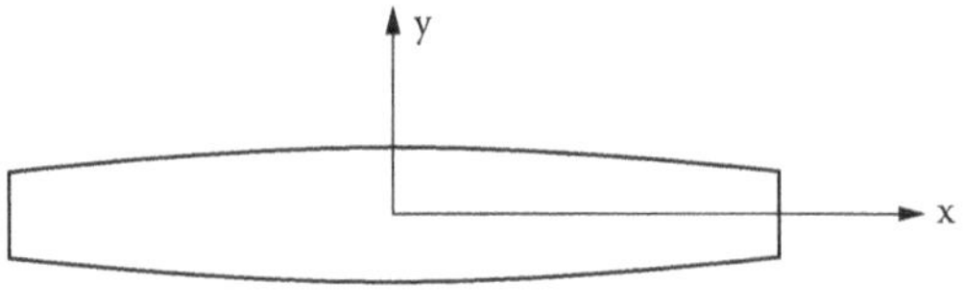

FIGURE 4.1
Variable section beam.

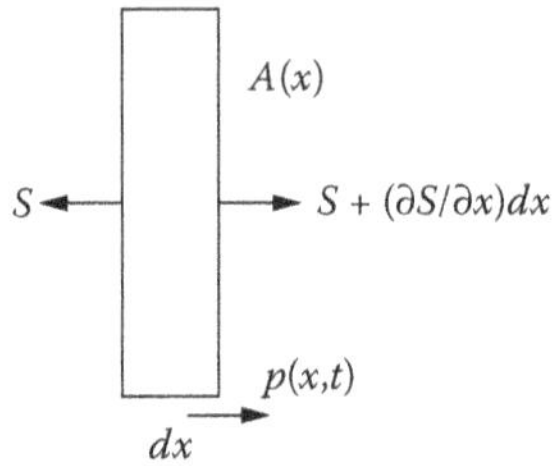

FIGURE 4.2
Longitudinal deflection element.

The equation of motion is then

$$S + (\partial S/\partial x)dx - S - \rho A dx(\partial^2 u/\partial t^2) = p(x,t)dx \tag{4.1}$$

The axial force S is

$$S(x) = \sigma(x)A(x) \tag{4.2}$$

where $\sigma(x)$ = longitudinal stress.

The longitudinal strain, $\varepsilon(x)$, is

$$\varepsilon(x) = \partial u/\partial x \tag{4.3}$$

and, in accordance with Hooke's law,

$$\sigma(x) = E(x)\varepsilon(x) \tag{4.4}$$

Thus,

$$\partial S/\partial x - \rho A(x)\partial^2 u/\partial t^2 = p(x,t) \tag{4.5}$$

and thus

$$\partial/\partial x[E(x)A(x)\partial u/\partial x] - \rho A(x)\partial^2 u/\partial t^2 = p(x,t) \tag{4.6}$$

4.2.2 Lateral Vibration

Consider a beam that has variable properties along its length. A differential element taken from any point along the beam is as shown in Figure 4.3. The beam has both bending and shearing properties, where M = bending moment, V = shearing force, and $p(x,t)$ = external loading per unit length.

Timoshenko[1] shows that, for beam bending, the strain $\varepsilon(x)$ is

$$\varepsilon(x) = y/r \tag{4.7}$$

where

y = distance from the middle plane of the beam to any point in the cross-section of the beam

r = radius of curvature of the middle surface due to bending

Using Hooke's law, the stress $\sigma(x)$ is therefore

$$\sigma(x) = E(x)y/r \tag{4.8}$$

where $E(x)$ = modulus of elasticity of the beam material, which can vary with x.

The bending moment, M, at location x is

$$M = \int \sigma(x)y\,dA = \int [E(x)/r]y^2 dA = E(x)I(x)/r \tag{4.9}$$

where $I(x)$ = moment of inertia of the beam as a function of x.

Figure 4.4 shows the bent beam. It is seen that

$$1/r = d\theta/ds, \text{ but } ds = dx \text{ and } \theta = dw/dx \; 1/r = d\theta/dx = d^2w/dx^2$$

Thus,

$$\mathrm{M} = \mathrm{E(x)I(x)d^2w/dx^2} \tag{4.10}$$

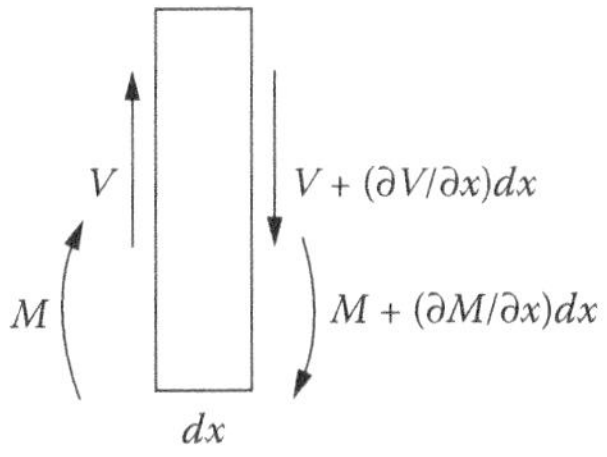

FIGURE 4.3
Lateral deflection element.

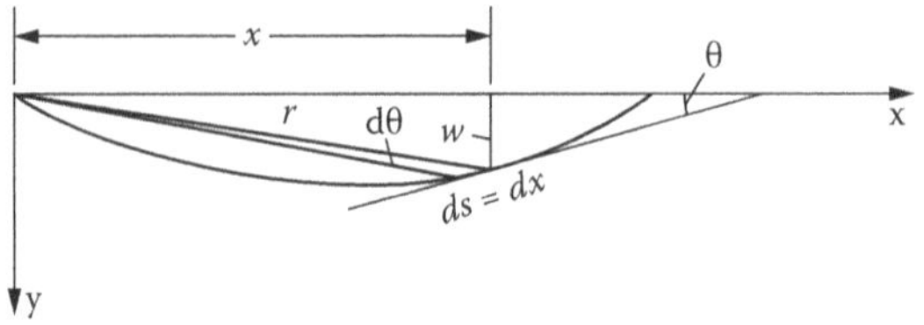

FIGURE 4.4
The bent beam. Slope at $x = dw/dx$.

These functional relationships are necessary for bending of beams and plates.

The equations of motion of the element dx are as follows:

$$M + (\partial M/\partial x)dx - M - Vdx = 0 \tag{4.11}$$

So

$$V = \partial M/\partial x \tag{4.12}$$

$$V + (\partial V/\partial x)dx - V = p(x,t)dx + \rho(x)A(x)d^2w/dt^2dx \tag{4.13}$$

Thus, the final equation of motion is

$$\partial^2/\partial x^2[(E(x)I(x)d^2w/dx^2] + \rho(x)A(x)d^2w/dt^2 = p(x,t) \tag{4.14}$$

Once the deflection is obtained for the longitudinal or lateral cases, the important parameters such as stresses, strains, velocities, accelerations, etc. can be obtained from the appropriate formulas.

The more complicated Timoshenko beam is discussed later. Solutions for some practical cases are given in this chapter.

4.3 Plates

All of the basic analysis for uniform (constant thickness) plates is contained in the treatise by Timoshenko.[2] The details for bending of unstiffened, stiffened, sandwich, and composite plates are discussed in Chapter 5. The use of plates in built-up structures is treated in Chapter 6. The analysis of plates has to include not only the bending and twisting of plates but also the stretching and shearing. The stretching and shearing of an element taken from the plate middle surface is shown in Figure 4.5.

Loading in stretching, shearing, bending, and twisting is shown in Figure 4.5. For the stretching and shearing, the potential energy, PE, is[2]

$$PE = (1/2)\ {}_A\!\int(N_x\varepsilon_x + N_y\varepsilon_y + N_{xy}Y_{xy})dA \tag{4.15}$$

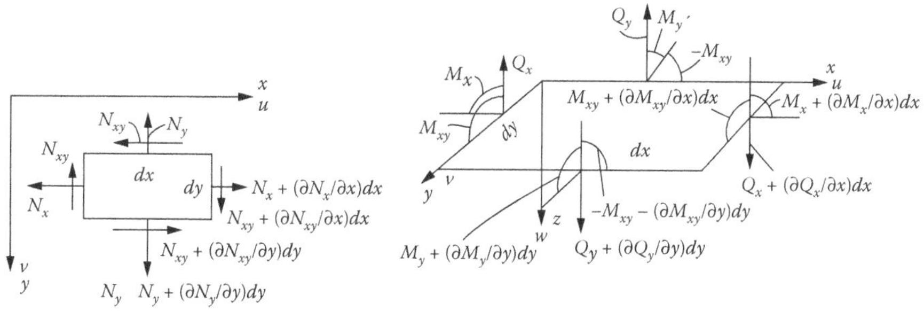

FIGURE 4.5
Loading on the element.

where

$$
\begin{aligned}
\varepsilon_x &= \partial u / \partial x + (1/2)(\partial w/\partial x)^2 \\
\varepsilon_y &= \partial v / \partial y + (1/2)(\partial w/\partial y)^2 \qquad (4.16) \\
Y_{xy} &= \partial u / \partial y + \partial v/\partial x + (\partial w/\partial x)(\partial w/\partial y)
\end{aligned}
$$

where
u = displacement in the x-direction
v = displacement in the y-direction
w = displacement in the z-direction
(u,v are the displacements in the middle plane and w is the lateral or bending displacement)

The strains are given in terms of forces by the following relations:

$$
\begin{aligned}
\varepsilon_x &= [1/E][(N_x/h) - \nu(N_y/h)] \\
\varepsilon_y &= [1/E][(N_Y/h) - \nu(N_X/h)] \qquad (4.17) \\
Y_{xy} &= N_{xy}/Gh
\end{aligned}
$$

The terms involving the deflection w are of a higher order and have been neglected in the analysis in Chapter 6. Thus, the simplified energy of stretching and shearing for a uniform constant thickness plate is

$$
PE = (1/2)_A \int\{[Eh/(1-\nu^2)][(\partial u/\partial x)^2 + (\partial v/\partial y)^2 + 2\nu(\partial u/\partial x)(\partial v/\partial y)] + Gh(\partial v/\partial x + \partial u/\partial y)^2\}dA \qquad (4.18)
$$

The energy of bending of a uniform plate is derived as follows:

Using the expression for the bending strain that was derived for the beam, it can be seen that

$$\varepsilon_x = -z\partial^2 w/\partial x^2 \tag{4.19}$$

where z is the distance from the middle surface of the plate to any point in the thickness of the plate and w is the lateral deflection. Also,

$$\begin{aligned} \varepsilon_y &= -z\partial^2 w/\partial y^2 \\ Y_{xy} &= -2z\partial^2 w/\partial x\partial y \quad \text{(Reference 2)} \end{aligned} \tag{4.20}$$

The stresses are obtained from Hooke's law:

$$\begin{aligned} \sigma_x &= [E/(1-\nu^2)](\varepsilon_x + \nu\varepsilon_y) \\ \sigma_y &= [E/(1-\nu^2)](\varepsilon_y + \nu\varepsilon_x) \end{aligned} \tag{4.21}$$

$$\tau_{xy} = G_{Y_{xy}} \qquad G = E/2(1+\nu)$$

The bending energy is

$$U_B = \int\int\int(\sigma_x\varepsilon_x/2 + \sigma_y\varepsilon_y/2 + \tau_{xy}Y_{xy}/2)dx\, dy\, dz \tag{4.22}$$

Thus,

$$\begin{aligned} U_B = (1/2)\int\int[Eh^3/12(1-\nu^2)][(\partial^2 w/\partial x^2)^2 + (\partial^2 w/\partial y^2)^2 \\ + 2\nu(\partial^2 w/\partial x^2)(\partial^2 w/\partial y^2) + 2(1-\nu)(\partial^2 w/\partial x\partial y)^2]dx\, dy \end{aligned} \tag{4.23}$$

The parts of the energy are the stretching and shearing given by Equation 4.18 and the bending and twisting given by Equation 4.23; relations for the stiffened, sandwich, and composite plates are derived in Chapter 5.

4.3.1 Differential Equations for Stretching and Shearing

The differential equations for the middle plane stresses can be written as follows:

$$\begin{aligned} \partial N_x/\partial x + \partial N_{xy}/\partial y + \rho h\partial^2 u/\partial t^2 &= p_x(x,y,t) \\ \partial N_{xy}/\partial x + \partial N_y/\partial y + \rho h\partial^2 v/\partial t^2 &= p_y(x,y,t) \end{aligned} \tag{4.24}$$

where p_x and p_y are the external loads in the x- and y-directions.

If we neglect the higher order terms in w in the strains, the equations become

$$Eh/(1-\nu^2)(\partial^2u/\partial x^2+\nu\partial^2v/\partial x\partial y)+Gh(\partial^2u/\partial x\partial y+\partial^2v/\partial y^2)+\rho h\partial^2u/\partial t^2=p_x$$
$$Eh/(1-\nu^2)(\partial^2v/\partial y^2+\nu\partial^2u/\partial x\partial y)+Gh(\partial^2u/\partial y^2+\partial^2v/\partial x\partial y)+\rho h\partial^2v/\partial t^2=p_y \quad (4.25)$$

4.3.2 Differential Equation for Bending and Twisting

The differential equation for bending and twisting starts with the moments and shearing forces (refer to Figure 4.5):

$$\partial Q_x/\partial x+\partial Q_y/\partial y+q=0 \qquad q=-f(x,y,t)+\rho h\ \partial^2w/\partial t^2$$

$f(x,y,t)$ = external load/unit area

h = thicknes of the uniform plate (4.26)

$$\partial M_{xy}/\partial x-\partial M_y/\partial y+Q_y=0$$
$$-\partial M_{xy}/\partial y+\partial M_x/\partial x-Q_x=0$$

Solving for Q_x and Q_y and substituting into the first equation, we obtain

$$\partial^2M_x/\partial x^2+\partial^2M_y/\partial y^2-2\ \partial^2M_{xy}/\partial x\partial y=-q \quad (4.27)$$

But,

$$M_x=\int_z\sigma_x z dz=-(z^3/3)E/(1-\nu^2)(\partial^2w/\partial x^2+\nu\partial^2w/\partial y^2)$$

upper limit on $z=+h/2$, lower limit on $z=-h/2$ (4.28)

$$M_y=\int_z\sigma_y z dz=-(z^3/3)E/(1-\nu^2)(\partial^2w/\partial y^2+\nu\partial^2w/\partial x^2)$$
$$M_{xy}=\int_z\tau_{xy}z dz=(z^3/3)2G\partial^2w/\partial x\partial y$$

Finally, we obtain

$$\partial^4w/\partial x^4+2\partial^4w/\partial x^2\partial y^2+\partial^4w/\partial y^4=q/D \quad (4.29)$$

where $D=Eh^3/(1-v^2)$.

4.4 Shells

The cylindrical shell is treated fully in Chapter 7. Chapter 6 analyzes shells that can be divided into a series of rectangular plates. The approximate theory is given in that chapter.

4.5 Three-Dimensional Variable Bodies

Consider a three-dimensional body that is composed of structure and fluid (including air), where the structure can be viscoelastic (i.e., the properties are complex numbers that include damping properties). The structure, fluid, and viscoelastic properties can vary with space. The equations of motion of such a body are given in Chapter 8 and a computer program is given for an approximate solution to some realistic problems. In this chapter, the basic equations will be derived.

The strains are given in terms of the stresses by Hooke's law as follows:

$$
\begin{aligned}
e_x &= \frac{1}{E}[\sigma_x - \mu(\sigma_y + \sigma_z)] + \alpha T & e_{xy} &= \frac{1+\mu}{E}\tau_{xy} \\
e_y &= \frac{1}{E}[\sigma_y - \mu(\sigma_x + \sigma_z)] + \alpha T & e_{yz} &= \frac{1+\mu}{E}\tau_{yz} \\
e_z &= \frac{1}{E}[\sigma_z - \mu(\sigma_x + \sigma_y)] + \alpha T & e_{zx} &= \frac{1+\mu}{E}\tau_{zx}
\end{aligned} \tag{4.30}
$$

$$
e_x = \frac{\partial u}{\partial x}, \quad e_y = \frac{\partial v}{\partial y}, \quad e_z = \frac{\partial w}{\partial z},
$$

$$
e_{xy} = \frac{\partial u}{\partial y} + \frac{\partial v}{\partial x}, \quad e_{xz} = \frac{\partial u}{\partial z} + \frac{\partial w}{\partial x}, \quad e_{yz} = \frac{\partial v}{\partial z} + \frac{\partial w}{\partial y}
$$

The more familiar constants E and G can be written in terms of λ and μ by the following relations:

$$
G = E/2(1+\mu) \qquad \lambda = \mu E/(1+\mu)(1-2\mu) \tag{4.31}
$$

The equations of motion will be derived in rectangular coordinates. Consider the three-dimensional elastic body as shown in Chapter 8.

The equations of motion in the three coordinate directions are as follows:

$$\begin{aligned}
&\frac{\partial \sigma_x}{\partial x}+\frac{\partial \tau_{yx}}{\partial y}+\frac{\partial \tau_{zx}}{\partial z}=-F_x+\rho\frac{\partial^2 u}{\partial t^2}\\
&\frac{\partial \tau_{xy}}{\partial x}+\frac{\partial \sigma_y}{\partial y}+\frac{\partial \tau_{zy}}{\partial z}=-F_y+\rho\frac{\partial^2 v}{\partial t^2}\\
&\frac{\partial \tau_{xz}}{\partial x}+\frac{\partial \tau_{yz}}{\partial y}+\frac{\partial \sigma_z}{\partial z}=-F_z+\rho\frac{\partial^2 w}{\partial t^2}
\end{aligned} \tag{4.32}$$

Substituting the expressions for the stresses, strains, and displacements given before and going through straight differentiation procedures—with the understanding that the properties E, λ, μ, and G can be complex numbers and spatially dependent—we will obtain the equations of motion shown in Chapter 8.

It is remarkable that, in principle, one can solve the resulting system with today's computers. If the system to be solved is composed of elastic and viscoelastic materials and surrounded by air and/or water, then the structural acoustic problem can be solved. The system can be divided into a series of boxes, which have the dimensions DX, DY, and DZ. For the structural part, we can have λ and G, which can vary with x, y, and z. For any parts that are water, $\lambda = \lambda$ (water) and $G = 0$. For parts that are air, $\lambda = \lambda$ (air) and $G = 0$. The λ and G values depend upon temperature (see Equation 4.3). The values of λ and G or E, μ are related by the previous expressions.

The displacements u,v,w will be complex algebraic linear equations if the expressions for the derivatives, which are given in Chapter 8, are employed (i.e., if the finite difference method is used to solve the problem of interest).

4.6 Analysis of Beams

4.6.1 Bernoulli–Euler and Timoshenko Beams

The equation for the Bernoulli–Euler constant section beam is shown in Equation 4.33 (see previous sections):

$$EI\partial^4 y/\partial x^4 + \rho A\partial^2 y/\partial t^2 + C\partial y/\partial t = P(x,t) \tag{4.33}$$

The equation for the constant section Timoshenko beam is shown in the following equation[4]:

$$EI\,\partial^4 y/\partial x^4 + \rho A\,\partial^2 y/\partial t^2$$
$$-(\rho I + EI\rho/kG)\partial^4 y/\partial x^2\partial t^2 + (\rho^2 I/kG)\partial^4 y/\partial t^4 + C\,\partial y/\partial t = P(x,t) \tag{4.34}$$

where
E = modulus of elasticity of the beam material
I = area moment of inertia about the center of gravity of the beam
ρ = mass density of the beam material
A = cross-sectional area of the beam
G = modulus of shear of the beam material
k = shearing constant
y = vertical deflection
$P(x,t)$ = force per unit length exciting the beam
C = factor that involves damping and water loading
w = width of beam

4.6.2 Green's Function for the Constant Section Timoshenko Beam

Let

$$y = \sum_{i=1}^{\infty} X_i Q_i e^{i\omega t} \tag{4.35}$$

Substituting into the Timoshenko equation, multiplying through by X_i, and integrating from 0 to L, the following relation for Q_i is obtained:

$$Q_i = F_0 X_i(x_o)/(ap - bp + cp + dp + jep) \tag{4.36}$$

where

$$ap = EI_o \int^L X_i X_i^{iv}\, dx$$

$$bp = \rho A \omega_o^2 \int^L X_i^2\, dx$$

$$cp = [(\rho I) + (EI\rho/kG)]\omega_o^2 \int^L X_i X_i^2\, dx$$

$$dp = (\rho^2 I/kG)\omega_o^4 \int^L X_i^2\, dx$$

$$ep = [C_d + \rho_o C_o w(\theta + j\chi)]\omega_o \int^L X_i^2\, dx$$

The program "TIMO. FOR," which is shown in Section 4.7 and can be downloaded from the publisher's website, contains the calculations of this Green's function program results are contained in Tables 4.1 and 4.2, which have the comparisons with results from Lueschen, Bergman, and McFarland.[5]

TABLE 4.1

Comparison of Green's Functions for a 20-in. Timoshenko Beam

$G(x,10) \times 10^{-5}$ $f = 100$ Hz				$G(x,10) \times 10^{-6}$ $f = 600$ Hz				$G(x,10) \times 10^{-7}$ $f = 2575$ Hz			
x	From Ref. 5	Timo Program	In Water	x	From Ref. 5	Timo Program	In Water	x	From Ref. 5	Timo Program	In Water
5.0	0.33	0.39	0.35	2.5	0.30	0.20	0.16	2.5	−0.60	−0.49	−0.62
7.5	0.73	0.80	0.72	5.0	0.70	0.60	0.48	5.0	−1.20	−1.22	−1.61
10	1.33	1.29	1.16	7.5	1.00	0.91	0.70	7.5	−1.53	−1.51	−2.08
12.5	2.00	1.80	1.62	10.0	0.70	0.86	0.56	10.0	−1.50	−1.56	−2.10
15.0	2.50	2.33	2.10	12.5	0	0.28	−0.05	12.5	−1.76	−1.96	−2.31
17.5	3.13	2.88	2.59	15.0	−1.10	−0.66	−1.03	15.0	−1.17	−1.77	−1.94
20.0	3.67	3.45	3.08	17.5	−2.40	−1.70	−2.10	17.5	0	−0.60	−0.72
				20.0	−3.70	−2.70	−3.10	20.0	1.50	0.60	−0.57

Note: Force = 1 lb at half length (10 in.).

TABLE 4.2

Comparison of Green's Functions for a 200-in. Timoshenko Beam

$G(x,100)$ $f = 1$ Hz				$G(x,100)$ $f = 10$ Hz				$G(x,100) \times 10^{-4}$ $f = 28$ Hz			
x	From Ref. 5	Timo Program	In Water	x	From Ref. 5	Timo Program	In Water	x	From Ref. 5	Timo Program	In Water
50	0.004	0.004	0.003	25	0.0038	0.0028	0.00021	25	−0.30	−0.38	−1.15
75	0.0087	0.0080	0.0061	50	0.0092	0.0078	0.00002	50	−0.67	−0.95	−3.31
100	0.0133	0.0130	0.0097	75	0.0138	0.0125	0.00095	75	−1.00	−1.18	−4.89
125	0.0193	0.018	0.013	100	0.0154	0.0139	0.00092	100	−1.25	−1.28	−5.36
150	0.024	0.023	0.017	125	0.0107	0.0101	0.00036	125	−1.83	−1.78	−5.09
175	0.031	0.029	0.021	150	0.0023	0.0049	−0.00053	150	−1.67	−1.66	−3.52
200	0.035	0.035	0.025	175	−0.0107	0.0029	−0.0016	175	0	−0.60	−0.90
				200	−0.023	0.0098	−0.0025	200	2.33	0.60	1.56

Note: Force = 1 lb at half length (100 in.).

4.6.3 Addition of Water Loading

Using the piston assumption as given in Chapter 3, the value of C is as follows:

$$C = C_d + \rho_o C_o(\theta + j\chi) \tag{4.37}$$

where
C_d is the damping constant
ρ_o is the density of the fluid
C_o is the sound velocity in the fluid
θ is the fluid damping function
χ is the fluid mass addition

These values are contained in the "Timo. For" program, and the results with water loading are shown in Tables 4.1 and 4.2.

4.6.4 Variable Section Beam

The Bernoulli–Euler variable section beam is as follows:

$$\partial^2/\partial x^2(EI\partial^2 y/\partial x^2) + \rho A\, \partial^2 y/\partial t^2 + C\, \partial y/\partial t = 0 \tag{4.38}$$

For relatively low frequencies, where the following condition is met,

$$\omega/C_o(A/\Pi)^{1/2} \ll 1 \tag{4.39}$$

The values of χ and θ are as follows[6]:

$$\chi = 0.48\ \omega/C_o(A)^{1/2}, \quad \theta = 0.13\ (\omega/C_o)^2 A \tag{4.40}$$

The values of the modal frequencies for a variable free–free beam are found by substituting

$$y_i = Q_i X_i(x) e_i^{i\omega t} \tag{4.41}$$

into the Bernoulli–Euler variable section equation, letting the external force equal to zero, substituting the value of the water loading, multiplying by X_i, and integrating from 0 to L. The result is the following equation for the natural frequencies:

$$\omega_i = \{[\int X_i \partial^2/\partial x^2(EI\ \partial^2 X_i/\partial x^2)\ dx]/[\int X_i^2 \rho A dx + 0.48 w \rho_o (wL)^{1/2} \int X_i^2 dx]\}^{1/2} \tag{4.42}$$

An example of a variable section beam that is of practical significance is the ship called the *Gopher Mariner,* which is discussed in McGoldrick et al.[3] The values of the weight and moment of inertia are obtained from Figures 4.6 and 4.7.[3] The computer program for calculating the natural frequencies is VARBEAM.FOR, which is shown in Section 4.7 and can be downloaded from the publisher's website.

The first three frequencies obtained from the simplified analysis and the comparison with the test values of the surface ship *Gopher Mariner* are shown in the following chart:

Frequencies (radians/sec)		
	Calculated	**Test Values**
First mode	5.1	8.7
Second mode	14	16.3
Third mode	26	22

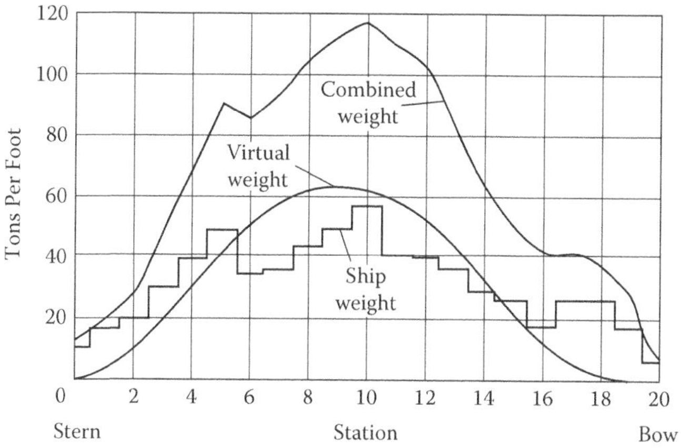

FIGURE 4.6
Weight curves for the *Gopher Mariner* (heavy displacement) used for calculation of vertical modes.

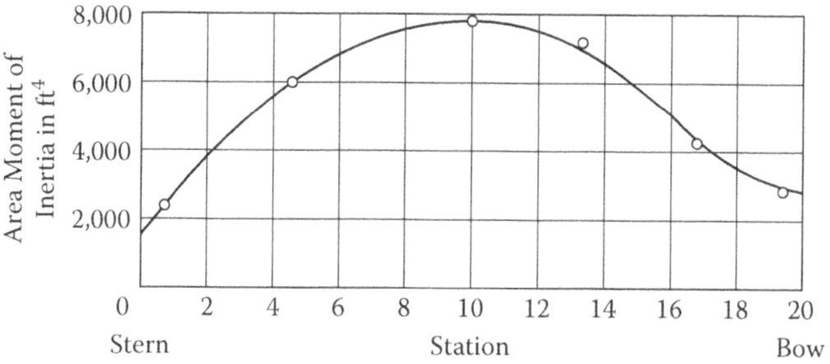

FIGURE 4.7
Area moment of inertia curve used for calculation of vertical modes of the *Gopher Mariner*.

4.6.5 Approximation to the Variable Section Beam Frequencies

If the average mass per unit length and the average moment of inertia are employed, the frequencies of the free-free variable beam are as follows:

$$\omega_i = [(2i+1)1.5708]^2/L^2 \; \{EI/[\rho A + 0.5\, w\rho_o(wL)^{1/2}]\}^{1/2} \qquad (4.43)$$

Using this approximation, the first three frequencies of the *Gopher Mariner* are shown in the following chart:

Frequencies (radians/sec)	
First mode	5
Second mode	13.5
Third mode	26.6

4.7 Computer Programs

```
                              TIMO. FOR
COMPLEX E1, ZIP, TEXP, DEN, Y
DI=1.33
F=1.0
RHO=7.28E-04
RHOW=9.32E-05
DKP=0.83
C0=60000.0
G=1.15E+07
CD=0.0
AR=4.0
E=3.0E+07
N1=1
N2=20
N3=1
IOMG1=1
IOMG2=2
IOMG3=3
IX1=10.0
IX2=400.0
IX3=10.
IT1=1
IT2=2
IT3=3
PI=3.1416
XL=200.0
X0=100.0
W=2.0
K2=20
A=DI*E
B=RHO*AR
C=(RHO*DI)+(E*DI*RHO/(DKP*G))
C   C=0
D=(RHO**2)*(DI/(DKP*G))
C   D=0
C   DO 902 IOMG=IOMG1, IOMG2, IOMG3
C   902 CONTINUE
DO 4000 IOMG=IOMG1, IOMG2, IOMG3
OMEG=IOMG*942.0
DO 3000 IX=IX1, IX2, IX3
X=IX*0.5
DO 2000 IT=IT1, IT2, IT3
T=0.
Y=0
DO 1000 N=N1, N2, N3
C   GO TO 20
```

```
10  IF(N.GT.1) GO TO 11
ALPHA=0.7340
BETNL=1.8751
GO TO 12
11  ALPHA=1.0
BETNL=(2.0*N-1.0)*1.5708
GO TO 13
12  BETN=BETNL/XL
ZR1=COSH(BETN*X)-COS(BETN*X)
ZR=ZR1-ALPHA*(SINH(BETN*X)-SIN(BETN*X))
ZR01=COSH(BETN*X0)-COS(BETN*X0)
ZR0=ZR01-ALPHA*(SINH(BETN*X0)-SIN(BETN*X0))
GO TO 14
13  BETN=BETNL/XL
ZR=EXP(-BETN*X)-COS(BETN*X)+SIN(BETN*X)
ZR0=EXP(-BETN*X0)-COS(BETN*X0)+SIN(BETN*X0)
GO TO 14
14  XI=2.0*ALPHA/BETN
                                TIMO
XIS=XL
XII=ALPHA*(BETN)*(2.0-ALPHA*BETNL)
XIV=(BETN**4)*XIS
GO TO 100
20  BETNL=N*PI
BETN=BETNL/XL
ZR=SIN(BETN*X)
C   PRINT *, BETNL
ZR0=SIN(BETN*X0)
XI=1.0
XIS=XL/2.0
XII=-(BETN**2)*XL/2.0
XIV=(BETN**4)*XL/2.0
GO TO 100
30  BETNL=(2.0*N+1)*1.5708
ALPHA=1.0
BETN=BETNL/XL
ZR=EXP(-BETN*X)+COS(BETN*X)-SIN(BETN*X)
ZR0=EXP(-BETN*X0)+COS(BETN*X0)-SIN(BETN*X0)
XIS=XL
XII=BETN*(2.0-BETNL)
XIV=(BETN**4)*XL
100 CHI=-0.48*(OMEG/C0)*SQRT(XL*W)
C   CHI=0.0
THETA=0.13*((OMEG/C0)**2)*XL*W
C   THETA=0.00
ZIP=CMPLX(THETA, CHI)
E1=CMPLX(CD+W*RHOW*C0*THETA, W*RHOW*C0*CHI)
C   GO TO 100
TEXP=CMPLX(COS(OMEG*T), SIN(OMEG*T))
```

```
AP=A*XIV
BP=-B*(OMEG**2)*XIS
CP=C*(OMEG**2)*XII
C   CP=0
DP=D*(OMEG**4)*XIS
C   DP=0
REP=(CD+W*RHOW*C0*THETA)*OMEG*XIS
C   REP=0
AEP=(W*RHOW*C0*CHI)*OMEG*XIS
C   AEP=0
DEN=CMPLX(AP+BP+CP+DP-AEP, REP)
C   PRINT *, N, AP, BP, CP, DP, AEP, REP
Y=Y+(F*ZR0*ZR*TEXP)/DEN
1000   CONTINUE
C   DO 923 IOMG=IOMG1, IOMG2, IOMG3
Y1=ABS(Y)
C   923 CONTINUE
C   PRINT *, OMEG, X, T, Y1
2000   CONTINUE
PRINT *, OMEG, X, T, Y
3000   CONTINUE
C   PRINT *, OMEG, X, T, Y1
4000   CONTINUE
C   PRINT *, OMEG, X, T, Y1
C   DO 962 IOMG=IOMG1, IOMG2, IOMG3
C   OMEG=IOMG*.1
C   Y1(IOMG)=ABS(Y(IOMG))
C   PRINT *, IOMG, OMEG, Y1(IOMG)
C   962 CONTINUE
C   DO 965 IOMG=IOMG1+1, IOMG2, IOMG3
C   OMEG=IOMG*.1
C   IF(Y1(IOMG-1)<Y1(IOMG)>Y1(IOMG+1)) THEN IB=IOMG
C   PRINT *, IB
```

TIMO

```
C   GO TO 965
C   216 PRINT *, OMEG, IOMG, Y1(IOMG-1), Y1(IOMG), Y1(IOMG+1)
C   965 CONTINUE
END
```

VARBEAM.FOR

```
DIMENSION XMU(30), DI(30), BX(30), BX1(30), BX2(30),
BX3(30), BX4(30)
DIMENSION XIP(30), XIPP(30), EXPP(30), X(30)
DX=315.0
E=3.0E+07
N=1
L=6300.0
W=400.0
RHOW=9.355E-05
```

```
BETN=(2*N+1.0)*1.5708/L
S1=0
S2=0
S3=0
DI(0)=5.0E+07
DI(1)=5.0E+07
DI(2)=8.3E+07
DI(3)=10.0E+07
DI(4)=12.4E+07
DI(5)=13.0E+07
DI(6)=14.0E+07
DI(7)=15.0E+07
DI(8)=16.0E+07
DI(9)=16.4E+07
DI(10)=16.6E+07
DI(11)=16.0E+07
DI(12)=15.5E+07
DI(13)=14.5E+07
DI(14)=13.5E+07
DI(15)=12.5E+07
DI(16)=10.5E+07
DI(17)=8.3E+07
DI(18)=7.5E+07
DI(19)=6.0E+07
DI(20)=5.8E+07
XMU(0)=6.0
XMU(1)=6.0
XMU(2)=8.6
XMU(3)=13.0
XMU(4)=17.2
XMU(5)=21.0
XMU(6)=16.0
XMU(7)=16.0
XMU(8)=18.0
XMU(9)=21.0
XMU(10)=25.0
XMU(11)=17.0
XMU(12)=17.0
XMU(13)=16.8
XMU(14)=12.0
XMU(15)=11.0
XMU(16)=8.6
XMU(17)=11.6
XMU(18)=11.6
XMU(19)=8.0
XMU(20)=4.0
X(0)=0
X(1)=315.0
X(2)=630.0
X(3)=945.0
```

```
X(4)=1260.0
X(5)=1575.0
X(6)=1890.0
X(7)=2205.0
X(8)=2520.0
```

VARBEAM

```
X(9)=2835.0
X(10)=3150.0
X(11)=3465.0
X(12)=3780.0
X(13)=4095.0
X(14)=4410.0
X(15)=4725.0
X(16)=5040.0
X(17)=5355.0
X(18)=5670.0
X(19)=5985.0
X(20)=6300.0
M1=2
M2=19
M3=1
DO 10 M=M1, M2, M3
A=BETN*X(M)
BX(M)=EXP(-A)+COS(A)-SIN(A)
C   PRINT *, BX(M)
BX1(M)=-BETN*(EXP(-A)+SIN(A)+COS(A))
BX2(M)=(BETN**2)*(EXP(-A)-COS(A)+SIN(A))
BX3(M)=(BETN**3)*(-EXP(-A)+SIN(A)+COS(A))
BX4(M)=(BETN**4)*BX(M)
XIP(M)=(DI(M+1)-DI(M-1))/(2.0*DX)
XIPP(M)=(DI(M+1)+DI(M-1)-2.0*DI(M))/(DX**2)
Q=E*DI(M)*BX4(M)
R=2.0*E*XIP(M)*BX3(M)
S=E*XIPP(M)*BX2(M)
EXPP(M)=Q+R+S
C   PRINT *, S
S1=S1+BX(M)*EXPP(M)*DX
C   PRINT *, S1
S2=S2+(BX(M)**2)*XMU(M)*DX
C   PRINT *, S2
S3=L
10 CONTINUE
S3=L
T=S3*0.5*W*RHOW*SQRT(W*L)
PRINT *, S1, S2, T
OMEGA = SQRT(S1/(S2+T))
PRINT *, N, OMEGA
END
```

References

1. Timoshenko, S. 1940. *Strength of materials, Part I,* 2nd ed. New York: D. Van Nostrand Co.
2. Timoshenko, S. 1940. *Theory of plates and shells.* New York: McGraw–Hill Book Co.
3. McGoldrick, R. T., Gleyzal, A. N., Hess, R. L., and Hess, G. K., Jr. 1951. Recent developments in the theory of ship vibration. David Taylor Model Basin report 739.
4. Timoshenko, S., Young, D. H., and Weaver, W., Jr. 1974. *Vibration problems in engineering,* 4th ed., 433. New York: John Wiley & Sons, Inc.
5. Lueschen, G. G. G., Bergman, L. A., and McFarland, D. M. 1996. Green's functions for uniform Timoshenko beams. *Journal of Sound and Vibration* 194 (1): 93–102.
6. Stenzel, H. 1952. Die akustische Strahlung der rechteckigen Kolbenmembrane. *Acoustica* 2:263–281.
7. Kinsler, L. E., and Frey, A. R. 1962. *Fundamentals of acoustics,* 2nd ed., 502. New York: John Wiley & Sons, Inc.

5

*Unstiffened, Stiffened, Sandwich, and Composite Plates**

5.1 Introduction

This chapter discusses the analysis of elastic plates. The plates can be with or without stiffeners, sandwich type, or made of composite materials. An approximate solution of simple plates with defined boundary conditions is presented in the first part of the chapter. The latter part of the chapter discusses stiffened and sandwich plates. Finally, plates made of composite materials are treated.

An important point to make at this point is that all the values for the characteristics of the stiffened, sandwich, and composite plates can be used in the next chapter to solve built-up shell problems.

5.2 Unstiffened Plates

The general theory of small vibrations of plates is given in references 1–7. The plate problem consists of finding a solution of the plate equation (5.1) that satisfies given conditions on the boundaries of the plate:

$$DV^4 w + \rho h \frac{\partial^2 w}{\partial t^2} = P(x, y, t) \cdots \tag{5.1}$$

where D, ρ, and h are the plate modulus, mass density, and thickness, respectively, and P is the load per unit area applied normal to the plate. If $\partial/\partial n$ denotes differentiation along the outward normal and $\partial/\partial s$ differentiation along the tangent, the boundary conditions for clamped, simply supported, and free plates can be written as follows (see Timoshenko[4]):

* See references 14 and 15.

(A) Clamped boundary:

$$w = 0 \quad \text{along } s$$
$$\frac{\partial w}{\partial n} = 0 \quad \text{along } s \tag{5.2}$$

(B) Simply supported boundary:

$$w = 0 \quad \text{along } s$$
$$M_n = 0 \quad \text{along } s \tag{5.3}$$

(C) Free boundary:

$$M_n = 0 \quad \text{along } s$$
$$V_n = Q_n - \frac{\partial M_{ns}}{\partial s} = 0 \text{ along } s \tag{5.4}$$

The terms w, $\partial w/\partial n$, M_n, V_n, Q_n, and M_{ns} are explained completely in Timoshenko.[4] When a plate is loaded by a lateral dynamic load, as considered here, the deflection w can be expressed as follows, provided stresses remain within the elastic limit:

$$w = \sum_{r=1}^{\infty} w_r q_r \cdots \tag{5.5}$$

where
w_r is the normal-mode function depending only on space and is taken as dimensionless
q_r is a function depending only on time and has the dimension of length
r corresponds to the mode number

If Equation 5.5 is substituted into Equation 5.1, the following expression is obtained:

$$\frac{D}{\rho h} \nabla^4 \sum_{r=1}^{\infty} w_r q_r + \sum_{r=1}^{\infty} w_r \frac{d^2 q_r}{dt^2} = \frac{P}{\rho h} \cdots \tag{5.6}$$

Now multiply both sides of this equation by one of the normal mode functions w_m and integrate over the area A_p of the plate:

$$\frac{D}{\rho h} \int_{A_p} w_m \left[\nabla^4 \sum_{r=1}^{\infty} w_r q_r \right] dA + \int_{A_p} w_m \left[\sum_{r=1}^{\infty} w_r \frac{d^2 q_r}{dt^2} \right] dA = \int_{A_p} \frac{P w_m}{\rho h} dA \cdots \tag{5.7}$$

The first term in this equation contains integrals of the form

$$\int_{A_p} w_m \nabla^4 w_r dA, \tag{5.8}$$

and the second term contains integrals of the form

$$\int_{A_p} w_m w_r dA. \tag{5.9}$$

Rayleigh[6] has demonstrated that

$$\int_{A_p} w_m w_r dA \tag{5.10}$$

is zero if r is not equal to m and if the boundary of the plate is wholly or partly clamped, simply supported, or free.

It can also be demonstrated that

$$\int_{A_p} w_m \nabla^4 w_r dA$$

is zero if r is not equal to m. If the plate is vibrating freely in one of its modes, then

$$w = w_r \sin p_r \ t\ldots \tag{5.11}$$

Substituting this into the differential equation for free vibrations, multiplying through by w_m, and integrating over the area of the plate, the following expression is obtained:

$$\frac{D}{\rho h}\int_{A_p} w_m \nabla^4 w_r dA = p_r^2 \int_{A_p} w_m w_r dA\ldots \tag{5.12}$$

Since $\int_{A_p} w_m w_r dA = 0,\ \int_{A_p} w_m \nabla^4 w_r dA = 0$

Now Equation 5.7 can be written as

$$\frac{D}{\rho h}\int_{A_p} w_m \nabla^4 w_m q_m dA + \int_{A_p} w_m^2 \frac{d^2 q_m}{dt^2} dA = \int_{A_p} \frac{P w_m dA}{\rho h}$$

or

$$\frac{d^2 q_m}{dt^2} + \frac{D}{\rho h}\frac{\int_{A_p} w_m \nabla^4 w_m dA}{\int_{A_p} w_m^2 dA} q_m = \frac{1}{\rho h}\frac{\int_{A_p} P w_m dA}{\int_{A_p} w_m^2 dA}\ldots \tag{5.13}$$

or, written in more compact notation,

$$\ddot{q}_m + p_m^2 q_m = A_m(t)\ldots \tag{5.14}$$

where the frequency of the mth mode of vibration is

$$p_m = \sqrt{\frac{D}{\rho h}}\left[\sqrt{\frac{\int_{A_p} w_m \nabla^4 w_m dA}{\int_{A_p} w_m^2 dA}}\right]\ldots \tag{5.15}$$

and

$$A_m(t) = \frac{1}{\rho h}\frac{\int_{A_p} P w_m dA}{\int_{A_p} w_m^2 dA}\ldots \tag{5.16}$$

The solution of Equation 5.14 is

$$q_m = C_1 \sin p_m t + C_2 \cos p_m t + \frac{1}{p_m}\int_0^t A_m(\tau)\sin p_m(t-\tau)d\tau\cdots \tag{5.17}$$

where C_1 and C_2 are constants of integration. The first two terms correspond to the free vibration and the third to the forced vibration. The complete expression for the deflection is then given as

$$w = \sum_{m=1}^{\infty} w_m\left[C_1\sin p_m t + C_2\cos p_m t + \frac{1}{p_m}\int_0^t A_m(\tau)\sin p_m(t-\tau)d\tau\right]\cdots \tag{5.18}$$

Assume that the plate is subjected to an impulsive load $G(x, y)\, f(t)$, where G is the space load distribution and $f(t)$ is the time load distribution, called the shape of the pulse. The initial conditions for this type of loading at $t = 0$ are

$$w = 0, \quad \frac{dw}{dt} = 0\cdots \tag{5.19}$$

Under these conditions, $C_1 = C_2 = 0$, so the deflection can be written as

$$w = \sum_{m=1}^{\infty} w_m\left[\frac{1}{p_m}\int_0^t A_m(\tau)\sin p_m(t-\tau)d\tau\right]\cdots \tag{5.20}$$

where

$$A_m(\tau) = \frac{1}{\rho h} f(\tau) \frac{\int_{A_p} G w_m dA}{\int_{A_p} w_m^2 dA} \cdots \tag{5.21}$$

Thus,

$$w = \sum_{m=1}^{\infty} w_m \left[\frac{1}{\rho h} \frac{\int_{A_p} G w_m dA}{p_m^2 \int_{A_p} w_m^2 dA} \right] \left[p_m \int_0^t f(\tau) \sin p_m (t-\tau) d\tau \right] \cdots \tag{5.22}$$

Both Frankland[8] and Salvadori[9] have studied the response of a single-degree-of-freedom system to pulses of various shapes and have evaluated the integral $p_m \int_0^t f(\tau) \sin p_m(t-\tau) d\tau$ for different pulses $f(\tau)$. If the load is gradually applied (i.e., the time of application of the load is much greater than the fundamental period), then $p_m \int_0^t f(\tau) \sin p_m(t-\tau) d\tau$ has a value of unity. Thus, the static deflection under a given load distribution G is

$$w_{\text{static}} = \sum_{m=1}^{\infty} w_m \left[\frac{1}{\rho h} \frac{\int_{A_p} G w_m dA}{p_m^2 \int_{A_p} w_m^2 dA} \right] \cdots \tag{5.23}$$

or, written more compactly,

$$w_{\text{static}} = \sum_m (w_{\text{static}})_{m\text{th mode}} \cdots \tag{5.24}$$

Since the stresses are space derivatives of the deflection, it may be concluded from Equations 5.22 and 5.23 that the total dynamic response, deflection or stress, is equal to the sum over all modes of the static deflection or stress in the mth mode multiplied by the response factor of the mth mode. The response factor is defined by the following equation:

$$R_m(t) = p_m \int_0^t f(\tau) \sin p_m (t-\tau) d\tau \cdots \tag{5.25}$$

The maximum value of the response factor is known as the load factor and is designated by L.

5.3 Rectangular Plates

To calculate the frequency and response (deflection or stress) of the plate, the normal modes of vibration must be known. Young[11] has employed the beam functions to approximate the normal modes of vibration of rectangular plates by using the Ritz method for his calculations. In the calculations that follow, the normal modes of the plate are approximated by the product of two beam functions (i.e., $w_{\mathrm{m}} = X_i Y_j$). (Note that the numbers i and j are connected with the mode—for example, for the first mode, $i = 1$, $j = 1$; for the second mode, $i = 1$, $j = 2$; etc. It should also be stated at this time that the product of the beam functions is not the exact expression for the modes of a plate since this combination does not, in general, satisfy the plate equation. However, it can be seen by comparison of frequencies, deflections, and stresses with those available in the literature that this assumption gives reasonable results.)

The functions X_i and Y_j depend on the boundary conditions of the plate. (Note that the product of beam functions constitutes an exact solution to the problem only in the case of a plate simply supported on all edges.) For example, if the plate has two opposite edges clamped, a third edge simply supported, and a fourth edge free, then X_i is the function corresponding to a clamped–clamped beam and Y_j to a simply supported free beam. Using this value of w_m and substituting it into Expressions 5.15 and 5.22, the following formulas are obtained for the frequency and deflection in the mth mode for plates with free, simply supported, or fixed edges:

$$p_{ij} = \sqrt{\frac{D}{\rho h}}\sqrt{\frac{(\beta_i)^4}{a^4} + \frac{(\beta_j)^4}{b^4} + \frac{2\int_0^a\int_0^b X_i X_i'' Y_j Y_j'' dxdy}{\int_0^a\int_0^b X_i^2 Y_j^2 dxdy}} \cdots \tag{5.26}$$

$$w_{ij} = X_i Y_j \frac{\int_0^a\int_0^b G(x,y) X_i Y_j dxdy}{\rho h p_m^2 \int_0^a\int_0^b (X_i Y_j)^2 dxdy} R_m \cdots \tag{5.27}$$

The expressions for the bending moment and bending stress are as follows:

$$(M_x)_{ij} = D\left(\frac{\partial^2 w}{\partial x^2} + \nu\frac{\partial^2 w}{\partial y^2}\right) = D(X_i'' Y_j + \nu X_i Y_j'') \frac{\int_0^a\int_0^b G(x,y) X_i Y_j dxdy}{\rho h p_m^2 \int_0^a\int_0^b (X_i Y_j)^2 dxdy} (R_m) \cdots \tag{5.28}$$

$$(M_y)_{ij} = D\left(\frac{\partial^2 w}{\partial y^2} + \nu\frac{\partial^2 w}{\partial x^2}\right) = D(X_i Y_j'' + \nu X_i'' Y_j)\frac{\int_0^a \int_0^b G(x,y) X_i Y_j \, dx dy}{\rho h p_m^2 \int_0^a \int_0^b (X_i Y_j)^2 \, dx dy}(R_m)\cdots \tag{5.29}$$

The bending stresses are given by

$$\sigma_x = \frac{6M_x}{h^2}$$
$$\sigma_y = \frac{6M_y}{h^2} \tag{5.30}$$

In the preceding expressions for the bending moments, the primes denote differentiation with respect to x or y (e.g., $X_i'' = d^2X_i/dx^2$, $Y_j'' = d^2Y_j/dy^2$).

The values of X_i and Y_j are given next for several types of beams. These expressions were obtained from Young and Felgar[10] and Young[11]:

$$X_i = \cosh\frac{\beta_i x}{a} - \cos\frac{\beta_i x}{a} - \alpha_i\left(\sinh\frac{\beta_i x}{a} - \sin\frac{\beta_i x}{a}\right)\cdots \tag{5.31}$$

$$Y_j = \cosh\frac{\beta_j y}{b} - \cos\frac{\beta_j y}{b} - \alpha_j\left(\sinh\frac{\beta_j y}{b} - \sin\frac{\beta_j y}{b}\right)\cdots \tag{5.32}$$

$$X_i = \sin\frac{\beta_i x}{a}\cdots \tag{5.33}$$

$$Y_j = \sin\frac{\beta_j y}{b}\cdots \tag{5.34}$$

Formulas in Expressions 5.25 and 5.26 hold for a beam with clamped–clamped, clamped-free, or clamped-supported edges. Formulas in Expressions 5.33 and 5.34 hold for a beam with supported–supported edges. Expressions 5.33 and 5.34 are the exact expressions for a plate with all of its edges simply supported. The values of β and α as well as the integrals $\int_0^a X_i X_i'' dx$, $\int_0^a X_i^2 dx$ and the values of X_i and X_i'' are contained in references 10–12. These references were used to compute the tables given in the chapter. The values used in the calculation are given in Table 5.1.

TABLE 5.1

Beam Function Values

Type of Beam	*i* or *j*	α_i, α_j	β_i, β_j	$b\int_0^b Y_j Y_j'' dy =$ or $a\int_0^a X_i X_i'' dx$	$\frac{\int_0^b Y_j^2 dy}{b} =$ or $\frac{\int_0^a X_i^2 dx}{a}$	Value of X_i[a]	Point at Which this Value of X_i Occurs	Value of $\frac{a^2}{\beta_i^2} X_i''$[a]	Point at Which This Value of $\frac{a^2}{\beta_i^2} X_i''$ Occurs	$\frac{\int_0^b Y_j dy}{b} =$ or $\frac{\int_0^a X_i dx}{a}$
Clamped–clamped	1	0.9825	4.7300	–12.3026	1	1.5882	$x = 0.5\,a$	2	$x = 0$	0.8309
	2	1.0008	7.8532	–46.0501	1	0	$x = 0.5\,a$	2	$x = 0$	0
	3	1.0000	10.9956	–98.9048	1	–1.4060	$x = 0.5\,a$	2	$x = 0$	0.3638
	5	1.0000	17.2788	–263.9980	1	1.4146	$x = 0.5\,a$	2	$x = 0$	0.2315
Supported–supported	1		3.1416	–4.9349	0.5	1	$x = 0.5\,a$	–1	$x = 0.5\,a$	0.6366
	2		6.2832	–19.7396	0.5	0	$x = 0.5\,a$	0	$x = 0.5\,a$	0
	3		9.4248	–44.4141	0.5	–1	$x = 0.5\,a$	1	$x = 0.5\,a$	0.2122
	5		15.7080	–123.3725	0.5	1	$x = 0.5\,a$	–1	$x = 0.5\,a$	0.1273
Clamped–free	1	0.7341	1.8751	0.8582	1	2	$x = a$	2	$x = 0$	0.7830
Clamped–supported	1	1.0008	3.9266	–11.5128	1	1.4449	$x = 0.5\,a$	2	$x = 0$	0.8599

Note: Here, a or b is the length of the beam, and the origin $x = 0$ or $y = 0$ is located at one end.

[a] The tabulations will remain valid if X_i is replaced by Y_j.

5.4 Addition of Water Loading

If water loading is added, then Equation 5.1 becomes

$$D\nabla^4 w + \rho h \frac{\partial^2 w}{\partial t^2} + [C_d + \rho_o c_o(\theta + i\chi)]\frac{\partial w}{\partial t} = P(x,y,t) \tag{5.35}$$

For calculation of the Green's function,

$$w(x,y,t) = \sum w_m(x,y)\bar{q}_m \varepsilon^{i\omega t} \tag{5.36}$$

Thus,

$$-\omega^2 \bar{q}_m + i\omega \frac{\bar{q}_m[C_d + \rho_o C_o(\theta + i\chi)]}{\rho h} + \bar{q}_m \frac{\int_{A_P} w_m \nabla^4 w_m dA_P}{\rho h \int_{A_P} w_m^2 dA_P} = \frac{w(x_o, y_o)P_o}{\rho h \int_{A_P} w_m^2 dA_P} \tag{5.37}$$

$$\bar{q}_m = \frac{\dfrac{w_m(x_o, y_o)P_o}{\rho h \int_{A_P} w_m^2 dA_P}}{\dfrac{\int_{A_P} w_m \nabla^4 w_m dA_P}{\rho h \int_{A_P} w_m^2 dA_P} - \omega^2 - \dfrac{\rho_o c_o \omega\chi}{\rho h} + \dfrac{i\omega[c_d + \rho_o c_o \theta]}{\rho h}} \tag{5.38}$$

where $w_m \equiv w_{ij} = X_i Y_j$ and where X_t and Y_j are beam functions.

5.5 Proposed Design Procedure for Simply Supported or Clamped Rectangular Plates under Uniform Impulsive Pressure

The recommended procedure for obtaining natural frequency, maximum deflection, and maximum stress in a plate subjected to uniform impact loading is as follows:

The *natural frequency* of the plate in Figure 5.1 is given by the general formula

$$F = \frac{K}{2\pi a^2}\sqrt{\frac{D}{\rho h}} \tag{5.39}$$

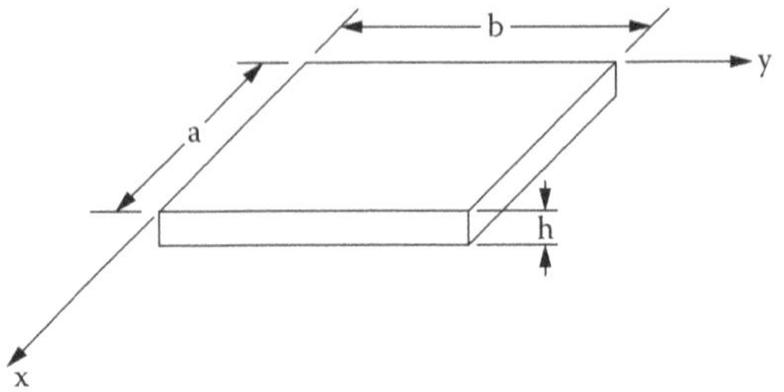

FIGURE 5.1
Sketch of rectangular plate.

where K is the frequency number (given in Table 5.2 for several modes of simply supported and clamped rectangular plates), which is a function of aspect ratio and mode, ρ is the mass per unit volume of the plate material, and

$$D = \frac{Eh^3}{12(1-\nu^2)}$$

where E is the modulus of elasticity and ν is Poisson's ratio for the plate material. For a steel plate with $E = 30 \times 10^6$ psi, $\nu = 0.80$,

$$\rho = \frac{0.284 \text{ lb-sec}^2}{386 \text{ in.}^4}, \quad F = 9780\frac{Kh}{a^2}\text{cps}$$

where h and a are measured in inches.

For an aluminum plate with $E = 10 \times 10^6$ psi, $\nu = 0.33$,

$$\rho = \frac{0.098 \text{ lb-sec}^2}{386 \text{ in.}^4}, \quad F = 9570\frac{Kh}{a^2}\text{cps}$$

where h and a are measured in inches.

The *maximum deflection* w_{max} in either a fixed or simply supported plate occurs at the center. An approximate upper limit for its value is computed as follows. (Warburton[2] has a procedure for calculating the natural frequencies of rectangular plates also based on the assumption that the modes can be represented by beam functions. He uses the Rayleigh method for his computations and gives an account of the limitations of assuming that the modes are composed of a single product of beam functions. However, he does not consider the calculation of deflections and stresses.) This assumes that the maximum contributions of the first four terms are additive:

TABLE 5.2

Frequencies, Deflections, and Stresses in Simply Supported and Clamped Rectangular Plates under Uniform Impulsive Load ($\nu = 0.3$)

	First Symmetrical Mode $w = X_1Y_1$			Scond Symmetrical Mode $w = X_1Y_3$			Third Symmetrical Mode $w = X_1Y_3$			Fourth Symmetrical Mode $w = X_3Y_1$			First Antisym Mode $w = X_1Y_2$	Second Antisym Mode $w = X_2Y_2$	Static Values: Deflection w; $w = y\frac{P_0a^4}{Eh^3}$; $y = A_1 - A_2 + A_3 - A_4$	Static Values: Stress σ; $\sigma = \delta\frac{6P_0a^2}{h^3}$; $\delta = B_1 - B_2^* + B_3 \pm B_4$
b/a	K	A_1	B_1	K	A_2	B_2	K	A_3	B_3	K	A_4	B_4	K	K	y	δ
Simply supported plate																
1.0	19.70	0.0454	0.0533	99.0	0.0006	0.0020	256.6	0.00085	0.0004	99.0	0.00060	0.0051	49.2	79.0	0.0443	0.0466
1.2	16.70	0.0632	0.0691	71.6	0.0011	0.0030	181.2	0.00011	0.0006	95.5	0.00064	0.0053	37.2	66.1	0.0616	0.0614
1.4	14.86	0.0796	0.0830	55.2	0.0019	0.0041	135.8	0.00019	0.0008	93.8	0.00067	0.0056	30.0	59.5	0.0772	0.0741
1.6	13.70	0.0939	0.0948	44.6	0.0030	0.0055	106.3	0.00032	0.0011	92.5	0.00068	0.0056	25.3	54.8	0.0905	0.0848
1.8	12.90	0.1061	0.1047	37.4	0.0042	0.0070	86.0	0.00048	0.0014	91.8	0.00069	0.0057	22.0	51.5	0.1017	0.0934
2.0	12.30	0.1163	0.1130	32.0	0.0057	0.0086	71.6	0.00069	0.0018	91.3	0.00078	0.0058	19.7	49.2	0.1106	0.1004
2.2	11.90	0.1248	0.1198	28.2	0.0074	0.0104	60.9	0.00096	0.0022	90.6	0.00071	0.0058	18.0	47.5	0.1177	0.1058
2.4	11.58	0.1319	0.1254	25.3	0.0092	0.0122	52.7	0.00127	0.0027	90.3	0.00072	0.0059	16.7	46.2	0.1233	0.1100
2.6	11.30	0.1379	0.1301	23.0	0.0111	0.0140	46.4	0.00165	0.0031	90.1	0.00072	0.0059	15.7	45.3	0.1277	0.1133
2.8	11.10	0.1429	0.1341	21.2	0.0131	0.0159	41.4	0.00207	0.0037	90.0	0.00072	0.0059	14.9	44.5	0.1312	0.1160
3.0	10.95	0.1472	0.1374	19.8	0.0151	0.0177	36.2	0.00255	0.0042	89.7	0.00073	0.0060	14.2	43.8	0.1339	0.1179
4.0	10.46	0.1609	0.1482	15.4	0.0247	0.0261	25.3	0.00554	0.0074	89.4	0.00074	0.0060	12.4	41.9	0.1410	0.1235
5.0	10.25	0.1680	0.1536	13.4	0.0326	0.0327	19.7	0.00908	0.0107	89.0	0.00074	0.0061	11.4	41.0	0.1437	0.1255
6.0	10.12	0.1720	0.1567	12.3	0.0386	0.0375	16.8	0.01258	0.0137	88.8	0.00074	0.0061	11.0	40.5	0.1452	0.1268
7.0	10.01	0.1750	0.1588	11.7	0.0431	0.0410	14.9	0.01601	0.0167	88.7	0.00074	0.0061	10.7	40.2	0.1472	0.1284
8.0	10.00	0.1760	0.1601	11.3	0.0465	0.0438	13.6	0.01905	0.0192	88.7	0.00074	0.0061	10.5	40.0	0.1478	0.1294
9.0	9.99	0.1773	0.1609	11.0	0.0491	0.0459	12.9	0.02124	0.0210	88.7	0.00074	0.0061	10.4	39.9	0.1483	0.1299
10.0	9.95	0.1780	0.1610	10.7	0.0509	0.0472	12.3	0.02326	0.0226	88.7	0.00074	0.0061	10.3	39.8	0.1496	0.1303

(Continued)

TABLE 5.2 (Continued)

Frequencies, Deflections, and Stresses in Simply Supported and Clamped Rectangular Plates under Uniform Impulsive Load ($\nu = 0.3$)

	First Symmetrical Mode $w = X_1Y_1$			Scond Symmetrical Mode $w = X_1Y_3$			Third Symmetrical Mode $w = X_1Y_3$			Fourth Symmetrical Mode $w = X_3Y_1$			First Antisym Mode $w = X_1Y_2$	Second Antisym Mode $w = X_2Y_2$	Static Values: Deflection w: $w = y\frac{P_0a^4}{Eh^3}$, $y = A_1 - A_2 + A_3 - A_4$	Static Values: Stress σ: $\sigma = \delta\frac{6P_0a^2}{h^3}$, $\delta = B_1 - B_2^* + B_3 \pm B_4$
b/a	K	A_1	B_1	K	A_2	B_2	K	A_3	B_3	K	A_4	B_4	K	K	y	δ
Clamped plate																
1.0	36.00	0.0146	0.0376	132.5	0.0004	0.0011	310.0	0.00005	0.00012	132.5	0.00042	0.0066	73.8	109.0	0.0138	0.0432
1.2	30.00	0.0199	0.0515	96.1	0.0008	0.0021	219.0	0.00010	0.00026	128.7	0.00045	0.0070	56.0	92.5	0.0188	0.0567
1.4	28.00	0.0242	0.0624	74.5	0.0013	0.0034	164.0	0.00017	0.00045	126.5	0.00046	0.0073	45.5	83.3	0.0226	0.0668
1.6	26.35	0.0273	0.0705	60.6	0.0020	0.0052	129.0	0.00028	0.00073	125.1	0.00047	0.0074	39.0	77.6	0.0251	0.0734
1.8	25.35	0.0296	0.0764	51.5	0.0028	0.0072	105.0	0.00043	0.00111	124.2	0.00048	0.0075	34.8	73.8	0.0260	0.0770
2.0	24.60	0.0313	0.0807	45.0	0.0036	0.0094	87.5	0.00061	0.00158	123.5	0.00048	0.0076	32.8	71.5	0.0278	0.0805
2.2	24.15	0.0325	0.0839	40.3	0.0045	0.0117	75.0	0.00084	0.00215	123.0	0.00049	0.0076	30.0	69.5	0.0284	0.0820
2.4	23.88	0.0334	0.0863	36.9	0.0054	0.0139	65.6	0.00109	0.00282	122.7	0.00049	0.0077	28.5	67.5	0.0286	0.0829
2.6	23.58	0.0341	0.0881	34.4	0.0062	0.0161	58.4	0.00130	0.00357	122.4	0.00049	0.0077	27.4	67.1	0.0288	0.0833
2.8	23.35	0.0347	0.0896	32.4	0.0070	0.0181	52.6	0.00169	0.00437	122.2	0.00049	0.0078	26.6	66.4	0.0289	0.0837
3.0	23.20	0.0352	0.0907	30.8	0.0078	0.0200	48.0	0.00203	0.00525	122.0	0.00049	0.0078	25.9	65.6	0.0289	0.0838
4.0	22.80	0.0364	0.0940	26.6	0.0104	0.0258	35.3	0.00376	0.00971	121.5	0.00050	0.0079	24.2	63.8	0.0293	0.0849
5.0	22.60	0.0370	0.0955	24.9	0.0119	0.0306	30.0	0.00523	0.01348	121.3	0.00050	0.0079	23.5	63.0	0.0298	0.0863
6.0	22.50	0.0373	0.0963	24.0	0.0127	0.0328	27.4	0.00627	0.01618	121.2	0.00050	0.0079	23.1	62.5	0.0304	0.0876
7.0	22.46	0.0375	0.0968	23.6	0.0133	0.0343	25.8	0.00706	0.01821	121.1	0.00050	0.0079	22.9	62.3	0.0308	0.0886
8.0	22.42	0.0376	0.0971	23.2	0.0136	0.0352	24.8	0.00761	0.01964	121.1	0.00050	0.0079	22.7	62.1	0.0311	0.0894
9.0	22.48	0.0377	0.0973	23.0	0.0139	0.0358	24.2	0.00794	0.02050	121.0	0.00050	0.0079	22.6	62.0	0.0312	0.0899
10.0	22.40	0.0378	0.0974	22.8	0.0141	0.0364	23.8	0.00822	0.02120	121.0	0.00050	0.0079	22.6	61.9	0.0314	0.0901
	1	2	3	4	5	6	7	8	9	10	11	12	13	14	15	16

Note: Only the symmetrical modes are excited in this case. *Positive sign in the case of the clamped plate and negative sign in the case of the simply supported plate.

$$w_{max} = \frac{P_0a^4}{Eh^3}[A_1L_1 + A_2L_2 + A_3L_3 + A_4L_4] \tag{5.40}$$

where A_1, A_2, A_3, and A_4 are tabulated in Table 5.2, and L_1, L_2, L_3, and L_4 are load factors for modes corresponding to A_1, A_2, A_3, and A_4 to be read off one of the curves given in Figure 5.2.

Column 15 of Table 5.2 gives an approximate value of the maximum deflection for static load application.

The *maximum tensile* or *compressive stress* occurs at the surfaces of a plate. Under uniformly distributed loading, it is located at the center of the plate in a simply supported plate and at the middle of the long side in a fixed plate. In both cases, the maximum tensile or compressive stress is in the direction of the short side. In a simply supported plate, there is a compressive stress on the top surface (where the load is applied) at the center; in a clamped plate, there is a tensile stress on the top surface at the edge. An upper limit for the maximum stress is computed as follows, assuming that the maximum contributions of the first four terms are additive:

$$\sigma_{max} = \frac{6}{h^2}P_0a^2[B_1L_1 + B_2L_2 + B_3L_3 + B_4L_4] \tag{5.41}$$

where B_1, B_2, B_3, and B_4 are tabulated in Table 5.2, and L_1, L_2, L_3, and L_4 are dynamic load factors to be read off one of the curves given in Figure 5.2.

Column 16 of Table 5.2 gives an approximate value of the maximum stress for static load application.

The plates considered in Table 5.2 are assumed to be fixed on all edges or simply supported on all edges. Tables 5.4 and 5.5 contain calculations for plates with other boundary conditions. The use of the first 14 columns of Table 5.2 is explained in the procedure, which is described at the beginning of Section 5.5. Columns 15 and 16 give the theoretical static response.

In the expressions for w_{max} and σ_{max} in the proposed design procedure, it was assumed that the vibratory modes were all in phase; therefore, the results obtained by using these expressions should always give maximum deflections or stresses higher than the values obtained by computing the response as a function of time. For example, if the load factor is assumed to be unity (which is the case for static loading), then the deflection or stress would be obtained, according to the proposed design procedure, by adding A_1, A_2, A_3, and A_4 in the case of deflection and B_1, B_2, B_3, and B_4 in the case of stress. The four-term approximation to the response should be adequate for pulses in which the first mode makes the primary contribution to the deflection and stress. It can be shown that the results obtained by adding these amplitudes are, in most cases, greater than the static

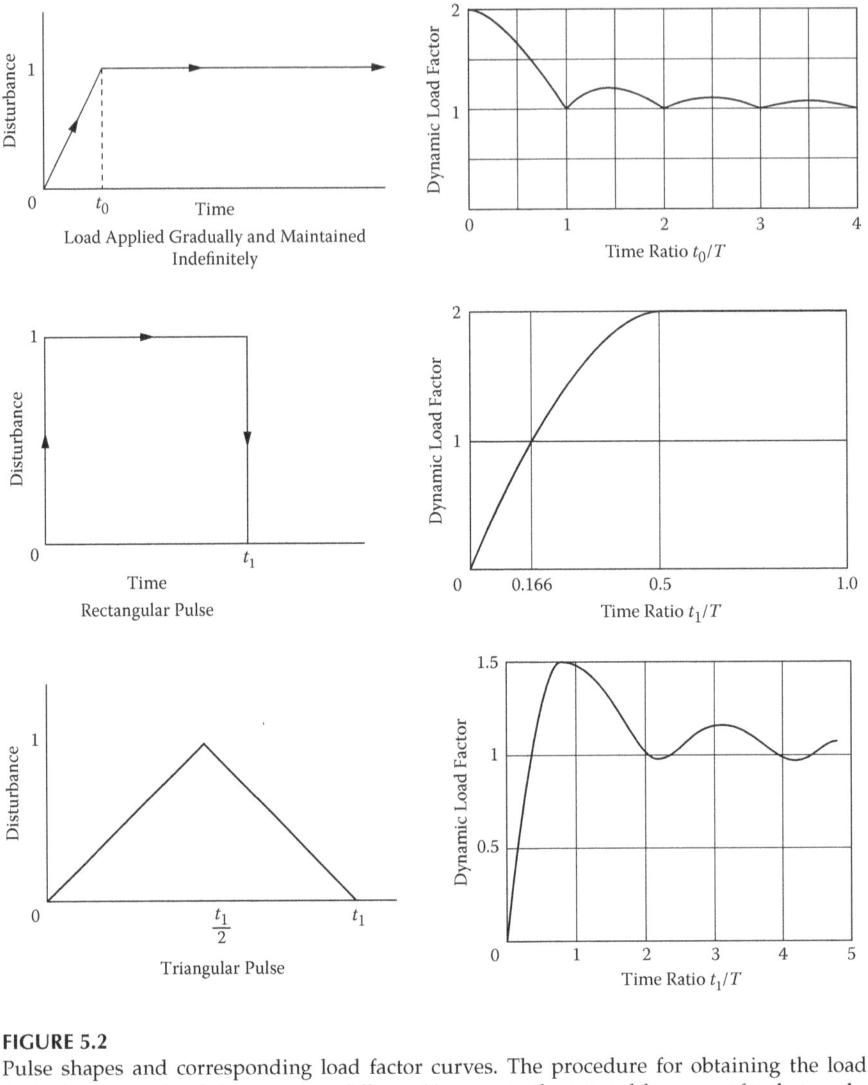

FIGURE 5.2
Pulse shapes and corresponding load factor curves. The procedure for obtaining the load factor for any mode of vibration is as follows: (1) compute the natural frequency for that mode; (2) invert this to obtain the period, T, for that mode; (3) for a given shape and duration of pulse, calculate t_1/T or t_0/T (where t_0 or t_1 is the time duration as shown in the pulse curves) and use the appropriate curve to determine the load factor for the mode of period T.

results obtained from the type of solution assumed here. For a few cases, however (e.g., the stress for a fixed plate with aspect ratios of 1, 1.2), the more exact solution given in Timoshenko[4] gives greater values than are obtained by summation of the first four terms given in columns 3, 6, 9, and 12 of Table 5.2.

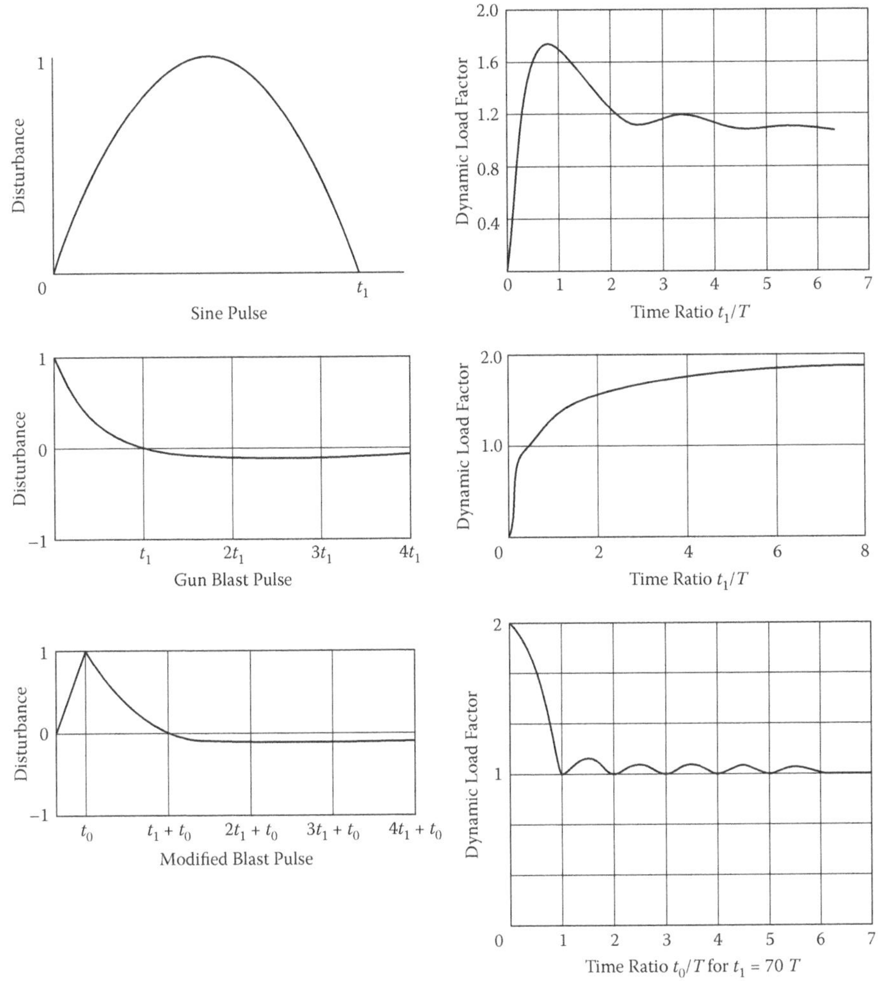

FIGURE 5.2 (Continued)
Pulse shapes and corresponding load factor curves. The procedure for obtaining the load factor for any mode of vibration is as follows: (1) compute the natural frequency for that mode); (2) invert this to obtain the period, T, for that mode; (3) for a given shape and duration of pulse, calculate t_1/T or t_0/T (where t_0 or t_1 is the time duration as shown in the pulse curves) and use the appropriate curve to determine the load factor for the mode of period T.

The following example is presented to illustrate the procedure.

Calculate the frequency of the first four symmetrical modes, the estimated maximum tensile stress, and the maximum deflection in a plate subjected to a triangular pulse load of 0.005 sec duration. The 24-by 48-by 0.5-in. plate is made of steel and has fixed edges. The magnitude of the load is 15 psi.

Natural Frequencies (for $b/a = 2$)

$$
\begin{aligned}
&\text{First Symmetrical Mode} && F_1 = \frac{9730\times 0.5}{(24)^2}\times 24.60 = 208 \text{ cps}\\
&\text{Second Symmetrical Mode} && F_2 = \frac{9730\times 0.5}{(24)^2}\times 45.00 = 380 \text{ cps}\\
&\text{Third Symmetrical Mode} && F_3 = \frac{9730\times 0.5}{(24)^2}\times 87.50 = 739 \text{ cps}\\
&\text{Fouth Symmetrical Mode} && F_4 = \frac{9730\times 0.5}{(24)^2}\times 123.00 = 1043 \text{ cps}
\end{aligned}
\tag{5.42}
$$

Natural Periods

$$
\begin{aligned}
&T_1 = 0.0048 \text{ sec} \quad \text{hence } \frac{t_1}{T_1} = \frac{0.0050}{0.0048} = 1.04\\
&T_2 = 0.0026 \text{ sec} \qquad \frac{t_1}{T_2} = \frac{0.0050}{0.0026} = 1.92\\
&T_3 = 0.0014 \text{ sec} \qquad \frac{t_1}{T_3} = \frac{0.0050}{0.0014} = 3.57\\
&T_4 = 0.0010 \text{ sec} \qquad \frac{t_1}{T_4} = \frac{0.0050}{0.0010} = 5.00
\end{aligned}
\tag{5.43}
$$

Load Factors
Referring to Figure 5.2, the following load factors are obtained:

$$
\begin{aligned}
L_1 &= 1.5\\
L_2 &= 1.0\\
L_3 &= 1.1\\
L_4 &= 1.1
\end{aligned}
\tag{5.44}
$$

Maximum Deflection

$$
\begin{aligned}
w_{max} &= \frac{P_0 a^4}{Eh^3}[0.0313\times 1.5 + 0.0036\times 1.0 + 0.00061\times 1.1 + 0.00048\times 1.1]\\
&= \frac{15\times (24)^4}{30\times 10^6\times (0.5)^3}\times 0.0517 = 0.0686 \text{ in.}
\end{aligned}
\tag{5.45}
$$

Maximum Stress

$$\sigma_{max} = \frac{6}{h^2} P_0 a^2 [0.0807 \times 1.5 + 0.0094 \times 1.0 + 0.00158 \times 1.1 + 0.0076 \times 1.1]$$

$$= \frac{6 \times 15 \times (24)^2}{(0.5)^2} \times 0.1405 = 29{,}134 \text{ psi}^* \quad (5.46)$$

5.5.1 Evaluation of Accuracy of Proposed Method

There were no experimental data or exact theoretical values with which to compare the dynamic deflections and stresses, so the static values are compared in Table 5.3 with those taken from Timoshenko.[4] The frequencies are compared with values obtained from Hearmon.[5]

Table 5.4 gives the formulas for the frequency numbers for the first mode for a number of plates, including the ones considered in Table 5.2. The results are compared with those obtained from Hearmon.[5]

Like Table 5.4, Table 5.5 gives results for other plates in addition to those given in Table 5.2. The cases considered in Table 5.2 are included as cases 1 and 2. The results in Table 5.5 were computed assuming that the total deflection and stress are represented by the first mode. Again, only the uniform load is considered, and results are compared with those taken from Timoshenko.[4]

5.5.2 Discussion

The methods developed provide the means for computing the responses of rectangular plates for a great number of cases. Table 5.2 may be utilized for the specific case of the simply supported or clamped plate under uniform impulsive load. Using this table, the designer will be able to obtain approximate frequencies, deflections, and stresses in a matter of minutes. A great amount of work plus a good knowledge of plate theory would be necessary if the designer desired to compute precise answers to his dynamic design problems. When finished, he would then probably apply a safety factor to his results. It is true that the methods given here do not give an exact solution to the problem; however, they permit an estimate of the responses and frequencies of plates under varying time and space load conditions. They will give answers to various problems that might otherwise remain unsolved, and these answers will at least be of the right order of magnitude.

* The very high value of stress is due to the fact that the duration of the load is almost equal to the fundamental period of the plate.

TABLE 5.3

Comparison of Frequency Parameter, Static Deflection, and Stress for Uniform Load ($\nu = 0.3$)

Sketch	b/a	Value of K (Table 5.1) First Mode	Second Mode	Third Mode	Fourth Mode	Value of K First Mode	Second Mode	Third Mode	Fourth Mode	Deflection γ (from Table 5.1)	γ (from Ref. 4)	Percent Difference	Stress δ (from Table 5.1)	δ (from Ref. 4)	Percent Difference
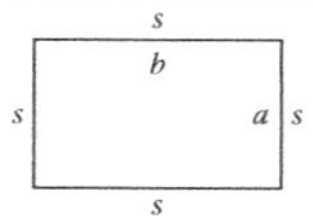	1.0	19.70	49.20	79.00	99.00	19.74	49.34	78.96	98.69	0.0443	0.0443	Insignificant	0.0466	0.0479	2.7
	1.2	16.70								0.0616	0.0616		0.0614	0.0626	1.9
	1.4	14.86								0.0772	0.0770		0.0741	0.0755	1.6
	1.6	13.70								0.0985	0.0906		0.0848	0.0862	1.6
Plate with all edges simply supported	1.8	12.90								0.1017	0.1017		0.0934	0.0948	1.5
	2.0	12.38				12.34				0.1106	0.1106		0.1004	0.1017	1.3
	3.0	10.95				10.97				0.1339	0.1336		0.1179	0.1189	0.8
	4.0	10.46								0.1410	0.1400	0.7	0.1235	0.1235	
	5.0	10.25								0.1437	0.1416	1.5	0.1255	0.1246	0.7
	10.0	9.95													
	—					9.87					0.1422			0.1258	
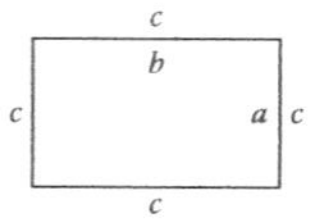	1.0	36.00	73.80	109.00	132.5[a]	35.99	73.41	108.3	132.3	0.0138	0.0138	0	0.0432	0.0513	15.8
	1.2	30.80								0.0188	0.0188	0	0.0567	0.0638	11.3
	1.4	28.00								0.0226	0.0226	0	0.0668	0.0726	8.0
	1.6	26.35								0.0251	0.0251	0	0.0734	0.0780	5.9
	1.8	25.35								0.0268	0.0267	Insignificant	0.0778	0.0812	4.2
Plate with all edges clamped	2.0	24.60				24.57				0.0278	0.0277		0.0805	0.0829	2.9
	3.0	23.20				23.19				0.0289			0.0838		
	4.0	22.60								0.0293			0.0848		
	5.0	22.60								0.0298			0.0863		
	10.0	22.40								0.0314			0.0901		
	—					22.37					0.0284			0.0830	

Note: The following formulas give the frequency, deflection, and stress respectively: $F = \frac{K}{2\pi a^2}\sqrt{\frac{D}{\rho h}}$; $w = \gamma \frac{P_0 a^4}{Eh^3}$; $\sigma = \delta \frac{6 P_0 a^2}{h^2}$. [a] The approximation in this case gives the correct value of frequency but not the correct nodal pattern.

TABLE 5.4

Comparison of First Mode Frequency Parameter for Plates with Combinations of Clamped and Simply Supported Edges ($\nu = 0.3$)

Case	Sketch[a]	Frequency Parameter	b/a	K (from Formula)	K (from Ref. 5)	K' (from Formula)	K' (from Ref. 5)
I	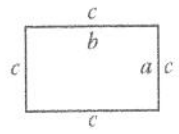	$K' = \sqrt{500(1+\eta^4)+303\eta^2}$	1.0	36.00	35.98	36.00	
			1.5	27.10	27.00	60.90	
			2.0	24.60	24.57	98.50	
			2.5	23.71	23.77	148.00	
			3.0	23.20	23.19	208.50	
			∞	22.37	22.37		
II	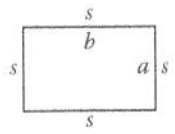	$K' = \sqrt{97.4(1+\eta^4)+194.8\eta^2}$	1.0	19.70	19.74	19.70	
			1.5	14.24	14.26	32.00	
			2.0	12.33	12.34	49.30	
			2.5	11.44	11.45	71.50	
			3.0	10.91	10.97	98.50	
			∞	9.87	9.87		
III	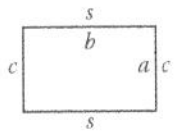	$K' = \sqrt{97.4\eta^4+500+243\eta^2}$	0			22.40	22.37
			0.333	207.00		23.00	22.99
			0.400	145.00		23.30	23.27
			0.500	95.20		23.80	23.82
			0.667	56.30		25.00	25.05
			1.000	29.00		29.00	28.95
			1.500	17.40	17.37	39.20	
			2.000	13.80	13.69	55.00	
			2.500	12.20	12.13	76.30	
			3.000	11.40	11.36	102.50	
			∞	9.87	9.87		
IV	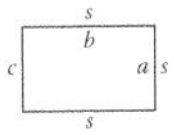	$K' = \sqrt{97.4\eta^4+238+227\eta^2}$	0			15.35	15.43
			0.333	145.80		16.20	16.26
			0.400	103.80		16.60	16.63
			0.500	69.20		17.30	17.33
			0.667	42.50		18.90	18.90
			1.000	23.70	23.65	23.70	23.65
			1.500	15.7	15.57	35.33	
			2.000	13.00	12.92	52.00	
			2.500	11.80	11.75	73.80	
			3.000	11.10	11.14	100.50	
			∞	9.87	9.87		
V	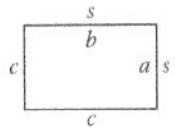	$K' = \sqrt{238(1+\eta^4)+265\eta^2}$	1.0	27.20		27.20	
			1.5	20.00		45.00	
			2.0	17.90		71.50	
			2.5	16.80		105.00	
			3.0	16.40		148.00	
			∞	15.35			

[a] C indicates clamped edge; S indicates simply supported edge.

$$F = \frac{K}{2\pi a^2}\sqrt{\frac{D}{\rho h}} = \frac{K'}{2\pi b^2}\sqrt{\frac{D}{\rho h}}; \eta = \frac{b}{a}; K = \frac{a^2}{b^2}K'$$

TABLE 5.5

Comparison of Static Deflection and Stress for Uniform Load Taking into Account Only First Term ($\nu = 0.3$)

Case	Sketch	b/a	Location of Deflection	Value $w = \gamma \frac{P_0 a^4}{Eh^3}$ γ	Value (from ref. 4) $w = \gamma \frac{P_0 a^4}{Eh^3}$ γ	Location of Stress	Value $\sigma = \frac{6}{h^2} \delta P_0 a^2$ δ	Value (from ref. 4) $\sigma = \frac{6}{h^2} \delta P_0 a^2$ δ
I	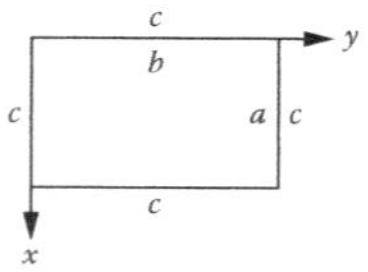	1.0		0.0146	0.0138		0.0376	0.0513
		1.5	$x = 0.5a$	0.0260	0.0240	$x = 0$	0.0669	0.0757
		2.0	$y = 0.5b$	0.0313	0.0277		0.0807	0.0829
		∞		0.0288	0.0284	$y = 0.5b$ (σ_x)	0.0740	0.0830
II	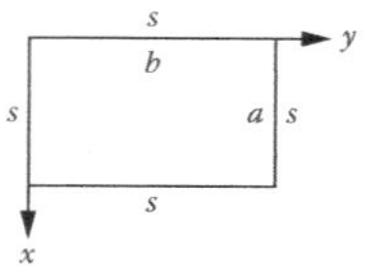	1.0		0.0454	0.0443		0.0533	0.0479
		1.5	$x = 0.5a$	0.0874	0.0843	$x = 0.5a$	0.0874	0.0812
		2.0	$y = 0.5b$	0.1163	0.1106	$y = 0.5b$	0.1130	0.1017
		∞		0.1430	0.1420	(σ_x)	0.1290	0.1250
III[a]	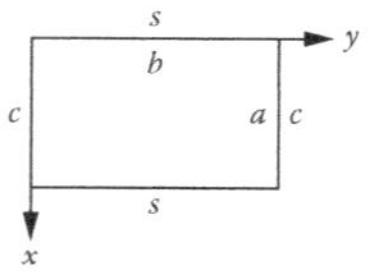	0.50		0.0020	0.0018		0.0208	0.0210
		0.67	$x = 0.5a$	0.0058	0.0053	$x = 0.5a$	0.0335	0.0370
		1.00		0.0218	0.0209		0.0563	0.0700
		1.50		0.0603	0.0582		0.0692	0.1050
		2.00	$y = 0.5b$	0.0969	0.0987	$y = 0$	0.0625	0.1190
		∞		0.1430	0.1422	(σ_y)	0.1290	0.1250

IV[a]	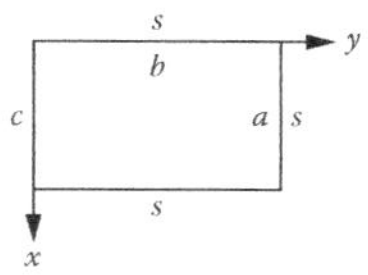	1.0	$x = 0.5a$	0.0307	0.0300	$x = 0.5a$	0.0600	0.0840
		1.5		0.0704	0.0700		0.0612	0.1120
		2.0	$y = 0.5b$	0.1022	0.1010	$y = 0$	0.0499	0.1220
		∞		0.1430	0.1420	(σ_y)	0.1290	0.1250
V	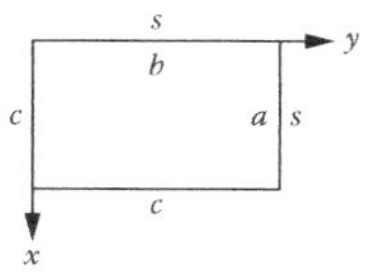	1.0		0.0227				
		1.5	$x = 0.5a$	0.0419				
		2.0	$y = 0.5b$	0.0528				
		∞		0.0570	0.0570			

[a] There is no consistency to the one-term approximations in calculating the stresses in these cases.

Examination of Tables 5.3–5.5 reveals the following information. As the aspect ratio b/a increases, the rate at which the frequency numbers as well as the deflection and stress amplitudes change becomes small rather quickly. This fact is very significant since it can be used to simplify many problems by using infinite aspect ratio formulas for plates with aspect ratios greater than 5.

The frequencies and deflections in clamped and simply supported plates under uniform static load are in agreement with the more accurate values in Timoshenko,[4] even though only four terms are considered in the series used to represent the deflection. (For aspect ratios greater than 4, Table 5.1 gives larger values for stress and deflection than the values for infinite aspect ratio predicted by the more exact theory.) The stresses, on the other hand, do not agree as well. For the simply supported plate, the stresses agree reasonably well; however, for the fixed plate, the error exceeds 10% for aspect ratios less than 1.2.

The one-term approximation for static values is reasonably good for deflection computations but must be used with caution for obtaining stresses. For large aspect ratios, however, the infinite aspect ratio formulas given should give good results. Since the contributions of the several modes are added, the magnitudes of deflection and stress are expected to be on the conservative side (overestimated) under the action of impulsive loads.

Although the accuracy of the method could only be compared for static load applications, it is believed that the same order of accuracy will be obtained for dynamic load applications.

The formulas given throughout the chapter hold only for small deflections (up to about 0.7 of the thickness of the plate) of flat plates. The more complicated shell theory must be used for plates with large initial curvature.

If the designer knows the type of loading to which the plate is subjected, the type of edge support, and that the stresses will always be within the elastic limit (or if the stresses should always be within the elastic limit), he can select a minimum thickness of plate with the information given in this chapter. The type of loading can be obtained from experimental records. Such records are now being created for ship plating. As far as the edge conditions of the plate are concerned, they could probably best be approximated by fixed edges if the plate is one panel of a larger plate containing stiffeners.

5.6 Cross-Stiffened and Sandwich Plates

The differential equation of motion for the deflection of an orthotropic plate—that is, one whose material has three planes of symmetry with respect to the elastic properties—can be written as[4]

$$D_x \frac{\partial^4 w}{\partial x^4} + 2H \frac{\partial^4 w}{\partial x^2 \partial y^2} + D_y \frac{\partial^4 w}{\partial y^4} + \mu_p \frac{\partial^2 w}{\partial t^2} + K \frac{\partial w}{\partial t} = P(x, y, t) \tag{5.47}$$

where

$$H = D_1 + 2D_{xy} \tag{5.48}$$

and D_x, D_y, D_1, and D_{xy} are the elastic constants for the stiffened plate, K^* is the structural viscous damping force per unit area per unit velocity, μ_p^{**} is the mass per unit area of the stiffened plate, and $P(x, y, t)$ is the external lateral pressure (load/unit area) applied to the plate.

Derivation of the method for calculating the elastic constants is given in Section 5.3.

Let the deflection w of the plate be expressed by the following summation of normal modes:

$$w = \sum_{r=1}^{\infty} w_r(x,y)\, q_r(t) \tag{5.49}$$

where
w_r is the dimensionless space-dependent normal mode function
q_r is a time-dependent function having the dimensions of length
r is the mode number

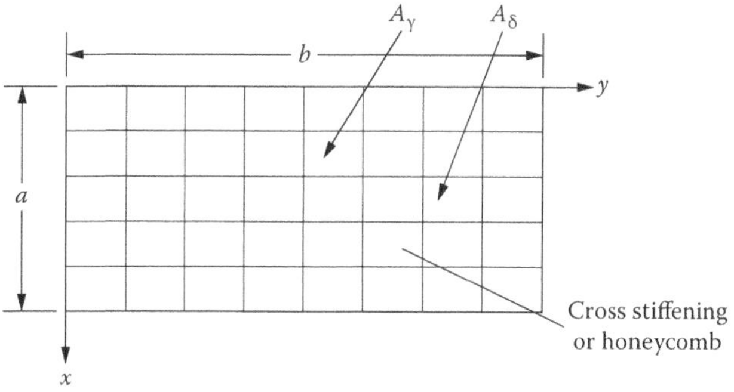

FIGURE 5.3
Cross-stiffened or honeycomb rectangular plate.

* K is calculated by assuming a given ratio of critical damping for the plate.

** μ_p is the total mass divided by the projected surface area ($a \times b$).

Substitution of Equation 5.41 into Equation 5.39 yields

$$\begin{aligned} &D_x \frac{\partial^4}{\partial x^4} \sum_{r=1}^{\infty} w_r q_r + 2H \frac{\partial^4}{\partial x^2 \partial y^2} \sum_{r=1}^{\infty} w_r q_r + D_y \frac{\partial^4}{\partial y^4} \sum_{r=1}^{\infty} w_r q_r \\ &+ \mu_p \frac{\partial^2}{\partial_t^2} \sum_{r=1}^{\infty} w_r q_r + k \frac{\partial}{\partial t} \sum_{r=1}^{\infty} w_r q_r = P(x,y,t) \end{aligned} \tag{5.50}$$

Integration of the product of Equation 5.42 and one of the normal mode functions w_m over the plate area A_p gives

$$\begin{aligned} &\int_{A_p} w_m \left[D_x \frac{\partial^4}{\partial x^4} \sum_{r=1}^{\infty} w_r q_r + 2H \frac{\partial^4}{\partial x^2 \partial y^2} \sum_{r=1}^{\infty} w_r q_r + D_y \frac{\partial^4}{\partial y^4} \sum_{r=1}^{\infty} w_r q_r \right] dA_p \\ &+ \mu_p \int_{A_p} w_m \frac{\partial^2}{\partial t^2} \sum_{r=1}^{\infty} w_r q_r dA_p + k \int_{A_p} w_m \frac{\partial}{\partial t} \sum_{r=1}^{\infty} w_r q_r dA_p = \int_{A_p} w_m P(x,y,t) dA_p \end{aligned} \tag{5.51}$$

If the plate boundary is wholly or partly clamped, simply supported, or free, then for $r \neq m$, the last two terms in the left-hand member are

$$\begin{aligned} &\mu_p \int_{A_p} w_m \frac{\partial^2}{\partial t^2} \sum_{r=1}^{\infty} w_r q_r dA_p + K \int_{A_p} w_m \frac{\partial}{\partial t} \sum_{r=1}^{\infty} w_r q_r dA_p \\ &= \mu_p \int_{A_p} \frac{\partial^2 q_r}{\partial t^2} \sum_{r=1}^{\infty} w_r w_m dA_p + K \int_{A_p} \frac{\partial q_r}{\partial t} \sum_{r=1}^{\infty} w_r w_m dA_p \end{aligned} \tag{5.52}$$

These are equal to zero because, for the stated boundary conditions, when $r \neq m$ (interchanging the integral and time derivatives in the right-hand members of Equation 5.44), the following orthogonality relations hold:

$$\int_{A_p} w_m w_r dA_p = 0 \tag{5.53}$$

If the plate is vibrating freely in one of its modes so that $P(x, y, t) = 0$, then, from Equation 5.44, the first term (consisting of the integral of three terms) in the left-hand member of Equation 5.43 is zero for $m \neq r$. Hence, with the forcing function $P(x, y, t)$ included, Equation 5.43 (including water loading via the piston approximation) can be written ($r = m$):

$$q_m \int_{A_p} w_m \left[D_x \frac{\partial^4 w_m}{\partial x^4} + 2H \frac{\partial^4 w_m}{\partial x^2 \partial y^2} + D_y \frac{\partial^4 w_m}{\partial y^4} \right] dA_p + \dot{q}_m \left[K \int_{A_p} w_m{}^2 dA_p \right]$$

$$+ \ddot{q}_m \left[\mu_p \int_{A_p} w_m{}^2 dA_p \right] = \int_{A_p} w_m P(x, y, t)\ dA_p - \dot{q}_m \int_{A_p} \rho_o c_o(\theta + i\chi) w_m{}^2 dA_p \tag{5.54}$$

or

$$\ddot{q}_m + \dot{q}_m \left\{ \frac{K' \int_{A_p} w_m{}^2 dA_p}{\mu_p \int_{A_p} w_m{}^2 dA_p} \right\} + q_m \left\{ \frac{\int_{A_p} w_m \left[D_x \frac{\partial^4 w_m}{\partial x^4} + 2H \frac{\partial^4 w_m}{\partial x^2 \partial y^2} + D_y \frac{\partial^4 w_m}{\partial y^4} \right]}{\mu_p \int_{A_p} w_m{}^2 dA_p} \right\}$$

$$= \frac{\int_{A_p} w_m P(x,y,t) dA_p}{\mu_p \int_{A_p} w_m{}^2 dA_p}$$

$$K' = K + \rho_o c_o(\theta + i\chi) \tag{5.55}$$

For free vibration, the equation takes the form of

$$\ddot{q}_m + 2n\dot{q}_m + p_m{}^2 \quad q_m = 0 \tag{5.56}$$

where

$$2n = \frac{K}{\mu_p} \tag{5.57}$$

The ratio of damping to critical damping can be written as

$$\frac{C}{C_C} = \frac{n}{p_m} \tag{5.58}$$

Thus,

$$K = 2\mu_p p_m \frac{C}{C_O} = \mu_p \frac{\delta}{\pi} p_m \tag{5.59}$$

To determine K, a value of C/C_C must be assumed, calculated, or measured and p_m must be calculated, where

$$p_m = \sqrt{\frac{\int_{A_p} w_m \left[D_x \frac{\partial^4 w_m}{\partial x^4} + 2H \frac{\partial^4 w}{\partial x^2 \partial y^2} + D_y \frac{\partial^4 w_m}{\partial y^4} \right] dA_p}{\mu_p \int_{A_p} w_m{}^2 dA_p}} \tag{5.60}$$

or the logarithmic decrement of ship plating can be used.

The effect of the water on the plate is introduced into the equations of motion through the function $P(x, y, t)$. The pressure $P(x, y, t)$ is divided into two parts:

1. the fluid pressure due to plate motion*
2. the forcing pressure $\bar{P}(x, y, t)$

Using the piston simplification for the water loading, the Green's function is calculated as follows:

$$q_m = \bar{q}_m \varepsilon^{iwt} \tag{5.61}$$

$$w(x,y,t) = \sum_m w_m(x,y)\, \bar{q}_m \varepsilon^{iwt} \tag{5.62}$$

$$-w^2 \bar{q}_m + \frac{iw\, \bar{q}_m [C_d + \rho_o c_o(\theta + i\chi)]}{\mu_p} + \bar{q}_m \frac{\int_{A_p} w_m \left[D_x \frac{\partial^4 w_m}{\partial x^4} + 2H \frac{\partial^4 w_m}{\partial x^2 \partial y^2} + D_y \frac{\partial^4 w_m}{\partial y^4} \right] a}{\mu_p \int_{A_p} w_m^2 dA_p}$$

$$= \frac{w_m(x_o, y_o)\rho_o}{\mu_p \int_{A_p} w_m^2 dA_p} \tag{5.63}$$

$$\bar{q}_m = \frac{\dfrac{w_m(x_o, y_o) p_o}{\mu_p \int_{A_p} w_m^2 dA_p}}{\dfrac{\int_{A_p} w_m \left[D_x \frac{\partial^4 w_m}{\partial x^4} + 2H \frac{\partial^4 w_m}{\partial x^2 \partial y^2} + D_y \frac{\partial^4 w_m}{\partial y^4} \right] dA_p}{\mu_p \int_{A_p} w_m^2 dA_p} - w^2 - \dfrac{\rho_o c_o w \chi}{\mu_p} + \dfrac{i\omega[c_d + \rho_o c_o \vartheta]}{\mu_p}} \tag{5.64}$$

where

$$w_m = w_{ij} = x_i\ y_j \tag{5.65}$$

x_i and y_j are beam functions

* The fluid pressure is opposite in direction but not equal to the forcing pressure.

5.7 Calculation of the Characteristics of Stiffened and Sandwich Plates

For convenience of reference, the definitions of the symbols are supplied in the following list:

E	Modulus of elasticity in tension and compression for plate material
G	Modulus of elasticity in shear for plate material
I	Area moment of inertia of cross-section of a beam
i_{w_x}, i_{w_y}	Moments of inertia of web about neutral planes of equivalent orthotropic plate
i_x', i_y', i_x'', i_y'', i_{xy}, i_{yx}	Defined by equations
M_x, M_y	Bending moments per unit length of section of a plate perpendicular to x- and y-axes, respectively
M_{xy}, M_{yx}	Twisting moments per unit length of section of a plate perpendicular to x- and y-axes, respectively
m_{w_x}, m_{w_y}	Unit bending moments in webs in two coordinate directions x and y, respectively
$r_{n'x}$, $r_{n'y}$	Scalar distance of centroid of near flange from neutral axis of bending in x- or y-direction, respectively
$r_{f'x}$, $r_{f'y}$	Scalar distance of centroid of far flange from neutral axis of bending in x- or y-direction, respectively
S_x, S_y	Distance between repeating sections in y- and x- directions, respectively
t_{nx}, t_{ny}	Thickness of near flanges (see Figures 5.3 and 5.4)
t_{fx}, t_{fy}	Thickness of far flanges (see Figures 5.3 and 5.4)
V	Total energy per unit area in a repeating section of a plate
V_O	Bending energy per unit area of an orthotropic plate
V_p	Bending and twisting energy per unit area in a repeating section of a plate

V_w	Energy in the webs per unit area for a repeating section of a plate
$w(x, y, t)$	Lateral deflection of a plate
x, y, z	Rectangular coordinate axes
$\varepsilon_{n'x}, \varepsilon_{n'y}$	Unit elongations for the near flange in the x- and y-directions, respectively
$\varepsilon_{f'x}, \varepsilon_{f'y}$	Unit elongations for the far flange in the x- and y-directions, respectively
ν	Poisson's ratio for plate material
σ_x, σ_y	Bending stresses in x- and y-directions, respectively
τ_{xy}, τ_{yx}	Shear stress in plate

Assuming that the webs of stiffeners act as simple beams in bending, the unit bending moments in the webs in the two coordinate directions x and y are

$$m_{w_x} = Ei_{w_x}\frac{\partial^2 w}{\partial x^2}; \quad m_{w_y} Ei_{w_y}\frac{\partial^2 w}{\partial y^2} \tag{5.66}$$

The energy in the webs of a single repeating section is then

$$\begin{aligned} V_w &= \tfrac{1}{2} m_{w_x}\left(\frac{\partial^2 w}{\partial x^2}\right) S_y + \tfrac{1}{2} m_{w_y}\left(\frac{\partial^2 w}{\partial y^2}\right) S_x \\ &= \tfrac{1}{2} Ei_{w_x}\left(\frac{\partial^2 w}{\partial x^2}\right) S_y + \tfrac{1}{2} Ei_{w_y}\left(\frac{\partial^2 w}{\partial y^2}\right) S_x \end{aligned} \tag{5.67}$$

5.7.1 Energy in the Plating

Referring to Figure 5.4, the basic expressions for the bending and twisting moments are (primes denote distance from neutral plane to center plane of flange)

$$M_x = \int_{r_{f'\lambda}-\frac{t_{fx}}{2}}^{r_{f'x}+\frac{t_{fx}}{2}} \sigma_x \, zdz + \int_{r_{n'x}-\frac{t_{nx}}{2}}^{r_{n'x}+\frac{t_{nx}}{2}} \sigma_x \, zdz \tag{5.68}$$

$$M_y = \int_{r_{f'y}-\frac{t_{fy}}{2}}^{r_{f'y}+\frac{t_{fy}}{2}} \sigma_y \, zdz + \int_{r_{n'y}-\frac{t_{ny}}{2}}^{r_{n'y}+\frac{t_{ny}}{2}} \sigma_y \, zdz \tag{5.69}$$

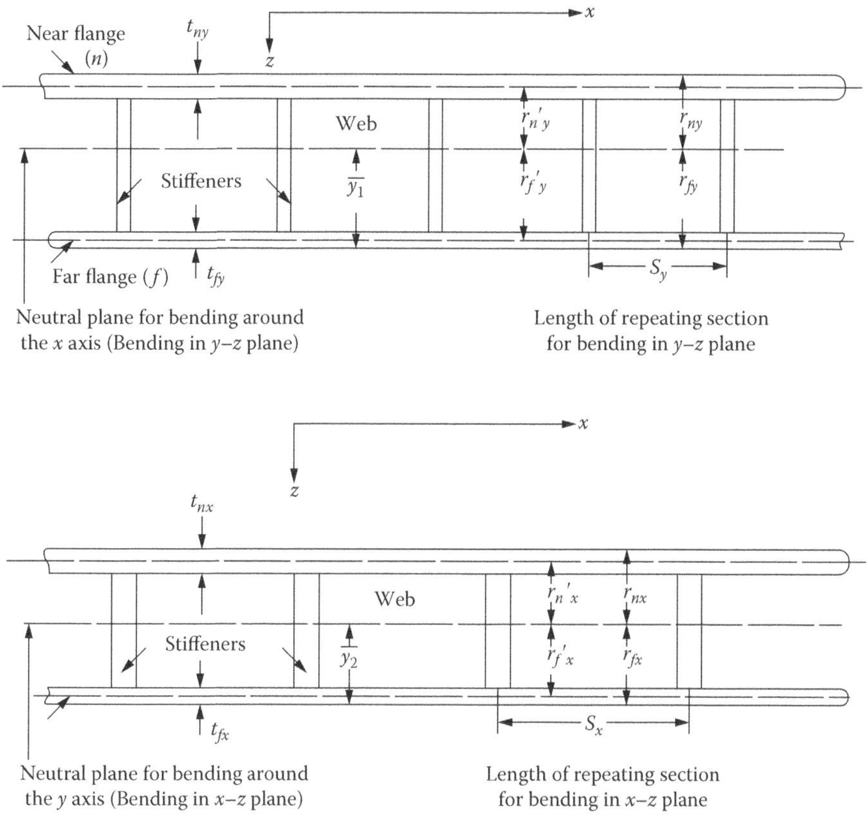

FIGURE 5.4
Cross-sectional views of stiffened plate. t_{nx}, t_{ny} is required to allow for different types of stiffeners with two different flange thicknesses running in orthogonal directions.

$$M_{xy} = \int_{r_{f'x}-\frac{t_{fx}}{2}}^{r_{f'x}+\frac{t_{fx}}{2}} \tau_{xy}\ zdz + \int_{r_{n'x}-\frac{t_{nx}}{2}}^{r_{n'x}+\frac{t_{nx}}{2}} \tau_{xy}\ zdz \tag{5.70}$$

$$M_{yx} = \int_{r_{f'y}-\frac{t_{fy}}{2}}^{r_{f'y}+\frac{t_{fy}}{2}} \tau_{yx}\ zdz + \int_{r_{n'y}-\frac{t_{ny}}{2}}^{r_{n'y}+\frac{t_{ny}}{2}} \tau_{yx}\ zdz \tag{5.71}$$

Substitute the following into Equation 5.59 (and analogously into Equation 5.60):*

* A linear distribution is assumed in the present report.

$$z = r_{n'x} \quad \text{or} \quad z = r_{f'x}$$

For the far flange (first term in the right-hand member),

$$\sigma_x = \frac{z}{r_{f'x}}\sigma_{f'x} = \frac{z}{r_{f'x}}\left[\frac{E}{1-\nu^2}(\varepsilon_{f'x}+\nu\varepsilon_{f'y})\right]$$
$$= \frac{z}{r_{f'x}}\left[\frac{E}{1-\nu^2}\left(r_{f'x}\frac{\partial^2 w}{\partial x^2}+\nu^r{}_{f'y}\frac{\partial^2 w}{\partial y^2}\right)\right] \tag{5.72}$$

For the near flange (second term in the right-hand member),

$$\sigma_x = \frac{z}{r_{n'x}}\sigma_{n'x} = \frac{z}{r_{n'x}}\left[\frac{E}{1-\nu^2}(\varepsilon_{n'x}+\nu\varepsilon_{n'y})\right]$$
$$= \frac{z}{r_{n'x}}\left[\frac{E}{1-\nu^2}\left(r_{n'x}\frac{\partial^2 w}{\partial x^2}+\nu r_{n'y}\frac{\partial^2 w}{\partial y^2}\right)\right] \tag{5.73}$$

Substitute the following into Equation 5.61 (and analogously into Equation 5.62):

For the far flange (first term in the right-hand member),

$$\tau_{xy} = \frac{z}{r_{f'x}}(\tau_{xy})_{f'x} = \frac{z}{r_{f'x}}\left[G(r_{f'x}+r_{f'y})\frac{\partial^2 w}{\partial x\partial y}\right] \tag{5.74}$$

For the near flange (second term in the right-hand member),

$$\tau_{xy} = \frac{z}{r_{n'x}}(\tau_{xy})_{n'x} = \frac{z}{r_{n'x}}\left[G(r_{n'x}+r_{n'y})\frac{\partial^2 w}{\partial x\partial y}\right] \tag{5.75}$$

After integration of Equations 5.59–5.62 and collection of terms $[\text{note } G = E/2(1+\nu)]$, this yields

$$M_x = \frac{E}{1-\nu^2}\left[i_{x'}\frac{\partial^2 w}{\partial x^2}+\nu i_x{}''\frac{\partial^2 w}{\partial y^2}\right] \tag{5.76}$$

$$M_y = \frac{E}{1-\nu^2}\left[i_{y'}\frac{\partial^2 w}{\partial y^2}+\nu i_y{}''\frac{\partial^2 w}{\partial x^2}\right] \tag{5.77}$$

$$M_{xy} = \frac{E}{2(1+\nu)}\left(i_{xy}\frac{\partial^2 w}{\partial x\partial y}\right) \tag{5.78}$$

$$M_{yx} = \frac{E}{2(1+\nu)}\left(i_{yx} \frac{\partial^2 w}{\partial x \partial y} \right) \tag{5.79}$$

where

$$i_x{}' = t_{fx} r_{f'x}^2 + t_{nx} r_{n'x}^2 + \frac{t_{fx}^3}{12} + \frac{t_{nx}^3}{12} \tag{5.80}$$

$$i_y{}' = t_{fy} r_{f'y}^2 + t_{ny} r_{n'y}^2 + \frac{t_{fy}^3}{12} + \frac{t_{ny}^3}{12} \tag{5.81}$$

$$i_x{}'' = t_{fx} r_{f'x} r_{f'y} + t_{nx} r'_{nx} r'_{ny} + \frac{r_{f'y}}{r_{f'x}} \frac{t_{f'x}^3}{12} + \frac{r_{n'y}}{r_{n'x}} \frac{t_{ny}^3}{12} \tag{5.82}$$

$$i_y{}'' = t_{fy} r'_{fx} r_{f'y} + t_{ny} r'_{nx} r'_{ny} + \frac{r_{f'x}}{r_{f'y}} \frac{t_{f'y}^3}{12} + \frac{r_{n'x}}{r_{n'x}} \frac{t_{ny}^3}{12} \tag{5.83}$$

$$\begin{aligned} i_{xy} &= t_{fx} r_{f'x}^2 + t_{nx} r_{n'x}^2 + \frac{t_{fx}^3}{12} + \frac{t_{nx}^3}{12} + t_{fx} r_{f'y} r_{f'x} \\ &\quad + t_{nx} r_{n'x} r_{n'y} + \frac{r_{f'y}}{r_{f'x}} \frac{t_{fx}^3}{12} + \frac{r_{n'y}}{r_{n'x}} \frac{t_{nx}^3}{12} \end{aligned} \tag{5.84}$$

$$\begin{aligned} i_{yx} &= t_{fy} r^2{}_{f'y} + t_{ny} r_{n'y}^2 + \frac{t_{fy}^3}{12} + \frac{t_{ny}^3}{12} + t_{fy} r_{f'x} r_{f'y} \\ &\quad + t_{ny} r_{n'y} r_{n'x} + \frac{r_{f'x}}{r_{f'y}} \frac{t_{f'y}^3}{12} + \frac{r_{n'x}}{r_{n'y}} \frac{t_{ny}^3}{12} \end{aligned} \tag{5.85}$$

For a plate of constant thickness t without stiffeners,

$$t_{fx} = t_{fy} = \frac{t}{2}; \quad t_{nx} = t_{ny} = \frac{t}{2}; \quad r_{n'x} = r_{n'y} = \frac{t}{4}; \quad r_{f'x} = r_{f'y} = \frac{t}{4}; \quad i_{w_x} = i_{w_y} = 0$$

By the use of these expressions, the effect of the stiffeners on the plate can be determined. The bending and twisting energy per unit area in a repeating section of plating is then

$$\begin{aligned} V_p &= \frac{1}{2}\left[M_x \frac{\partial^2 w}{\partial x^2} + M_y \frac{\partial^2 w}{\partial y^2} + M_{xy} \frac{\partial^2 w}{\partial x \partial y} M_{yx} \frac{\partial^2 w}{\partial x \partial y} \right] \\ &= \frac{1}{2}\left[\frac{E}{1-\nu^2}\left(i_{x'} \frac{\partial^2 w}{\partial x^2} + \nu i_x{}'' \frac{\partial^2 w}{\partial y^2} \right) \frac{\partial^2 w}{\partial x^2} \right. \end{aligned} \tag{5.86}$$

$$+\frac{E}{1-\nu^2}\left(i_{y'}\frac{\partial^2 w}{\partial y^2}+\nu i_y{}''\frac{\partial^2 w}{\partial x^2}\right)\frac{\partial^2 w}{\partial y^2}$$
$$+\frac{1}{2}\frac{E}{1+\nu}(i_{xy}+i_{yx})\left(\frac{\partial^2 w}{\partial x\partial y}\right)^2\Bigg] \tag{5.87}$$

The total energy per unit area in a repeating section of the plate obtained by summating Equations 5.58 and 5.78 is

$$V=V_p+V_w=\frac{1}{2}\left[\frac{Ei_{x'}}{1-\nu^2}+\frac{Ei_{wx}}{S_x}\right]\left(\frac{\partial^2 w}{\partial x^2}\right)$$
$$+\left(\frac{Ei_{y'}}{1-\nu^2}+\frac{Ei_{wy}}{S_y}\right)\left(\frac{\partial^2 w}{\partial y^2}\right)^2$$
$$+\left(\frac{\nu E}{1-\nu^2}i_x{}''+\frac{\nu E}{1-\nu^2}i_y{}''\right)\left(\frac{\partial^2 w}{\partial x^2}\right)\left(\frac{\partial^2 w}{\partial y^2}\right) \tag{5.88}$$
$$+\frac{1}{2}\frac{E}{1+\nu}(i_{xy}+i_{yx})\left(\frac{\partial^2 w}{\partial x\partial y}\right)^2\Bigg]$$

This is equated to the bending energy per unit area of an orthotropic plate, which may be written as

$$V_o=\frac{1}{2}\left[D_x\left(\frac{\partial^2 w}{\partial x^2}\right)+2D_1\frac{\partial^2 w}{\partial x^2}\frac{\partial^2 w}{\partial y^2}\right.$$
$$\left.+D_y\left(\frac{\partial^4 w}{\partial y^2}\right)^2+4D_{xy}\left(\frac{\partial^2 w}{\partial x\partial y}\right)^2\right] \tag{5.89}$$

to give the following expressions for the orthotropic constants:

$$D_x=\frac{E}{1-\nu^2}\left[i_x{}'+\frac{1-\nu^2}{S_x}i_{w_x}\right] \tag{5.90}$$

$$D_y=\frac{E}{1-\nu^2}\left[i_y{}'+\frac{1-\nu^2}{S_y}i_{w_y}\right] \tag{5.91}$$

$$D_1=\frac{\nu E}{1-\nu^2}\left[i_x{}''+i_y{}''\right] \tag{5.92}$$

$$D_{xy}=\frac{1}{2}\frac{E}{4(1+\nu)}\left[i_{xy}+i_{yx}\right] \tag{5.93}$$

5.8 Composite Plates[4,16]

In the case of plates made of composite materials, if the axes of the plate correspond to the material axes, the differential equation of the plate is[16]

$$D_{11}(\partial^4 w / \partial x^4) + 2D_{33}(\partial^4 w / \partial x^2 \partial y) + D_{22}(\partial^4 w / \partial y^4) + \mu_p \partial^2 w / \partial t^2 + K\partial w / \partial t = P(x, y, t) \quad (5.94)$$

where[17]

$$D_{33} = D_{12} + 2D_{33},\ D_{11} = E_1 h^3/12(1 - \nu_{12}^2),$$
$$D_{22} = E_2 h^3/12(1 - \nu_{12}^2) \quad (5.95)$$

For a uniform isotropic plate,

$$D_{11} = D_{22} = D_{33} = D = Eh^3/12(1 - \nu^2)$$

Eckold[16] gives some values for these coefficients:

$P(x, y, t)$ = external load per unit area
μ_p = mass per unit area
K = damping constant
h = thickness of plate

5.9 Plate Applications

The work on plates shown in this chapter can be applied to fixed or free plates or plates with any practical boundary conditions. The beam function method will be a useful tool for representing the modes.

The most general application will be for the development shown in the next chapter, where a built-up shell structure can be divided into a series of plates. Each plate has the characteristics developed in Chapter 3. Thus, a shell made up of many of these plates with different characteristics can be analyzed.

References

1. Bureau of Ships letter S11-3(442-443-330) of December 11, 1947, to David Taylor Model Basin.
2. Warburton, G. B. 1954. The vibration of rectangular plates. *Chartered Mechanical Engineer* January: 22.

3. Huffington, N. J., Jr., and Hoppmann, W. H., II. 1952. Blast loading of walls. Operational Research Office, Johns Hopkins University, technical memorandum ORO-T-199.
4. Timoshenko, S. 1940. *The theory of plates and shells.* New York: McGraw–Hill Book Co.
5. Hearmon, R. F. S. 1952. The frequency of vibration of rectangular isotropic plates. *Journal of Applied Mechanics* September: 402.
6. Rayleigh, J. W. S. 1945. *The theory of sound,* vol. I, chap. X. New York: Dover Publications.
7. Courant, R., and Hilbert, D. 1953. *Methods of mathematical physics,* vol. 1, chap. V. New York: Interscience Publishers Inc.
8. Frankland, J. M. 1942. Effects of impact on simple elastic structures. David Taylor Model Basin report 481.
9. Salvadori, M. G. 1947. A mathematical treatment of the generalized Bertz impact of a mass on a simply supported beam. *Welding Journal* July: 426.
10. Young, D., and Felgar, R. P. 1949. Tables of characteristic functions representing the normal modes of vibration of a beam, Engineering Research Series no. 44, University of Texas (July 1, 1949).
11. Young, D. 1950. Vibration of rectangular plates by the Ritz method. *Journal of Applied Mechanics* December: 448.
12. Felgar, R. P. 1950. Formulas for integrals containing characteristic functions of a vibrating beam. University of Texas circular no. 14.
13. Timoshenko, S. 1948. *Vibration problems in engineering,* 2nd ed. New York: D. Van Nostrand Co., Inc.
14. Greenspon, J. E. 1955. Stresses and deflections in flat rectangular plates under dynamic lateral loads based on linear theory. David Taylor Model Basin report 774.
15. Leibowitz, R. C., and Greenspon, J. E. 1964. A method for predicting the plate-hull girder response of a ship incident to slam. David Taylor Model Basin report 1706.
16. Eckold, G. 1994. *Design and manufacture of composite structures.* New York: McGraw–Hill, Inc.

6

A Simplified Procedure for Built-up Structures*

6.1 General Considerations

It will be assumed that the built-up shell can be divided into a series of cross-stiffened plates as shown in Figure 6.1. Displacements of the corners of these elements will be referred to as a fixed system of axes OXYZ as shown in the figure. These displacements will be denoted by UVW. The shell will be broken into these elements by a set of orthogonal lines and the elements will be numbered by the intersection of these lines as shown in the figure.

The free-body diagram of a representative plate element is shown in Figure 6.2. The local axes of the element are described by the lowercase letters x, y, and z. Let the direction cosines of the x-, y-, and z-axes with respect to the fixed XYZ axes be denoted as in Table 6.1.

Then, if the displacements of any point relative to the local axes are denoted by u, v, and w, the following relations will hold:

$$
\begin{aligned}
u &= \lambda_1 U + \mu_1 V + p_1 W \\
v &= \lambda_2 U + \mu_2 V + p_2 W \\
w &= \lambda_3 U + \mu_3 V + p_3 W
\end{aligned} \tag{6.1}
$$

6.2 Potential Energy and Stiffness Coefficients

The potential energy of deformation of the cross-stiffened plate consists of the membrane energy of stretching and shearing plus the energy of bending

* From Greenspon, J. E. 1963. Theoretical developments in the vibrations of hulls. *Journal of Ship Research* 6 (4). Reprinted with permission from the Society of Naval Architects and Marine Engineers (SNAME).

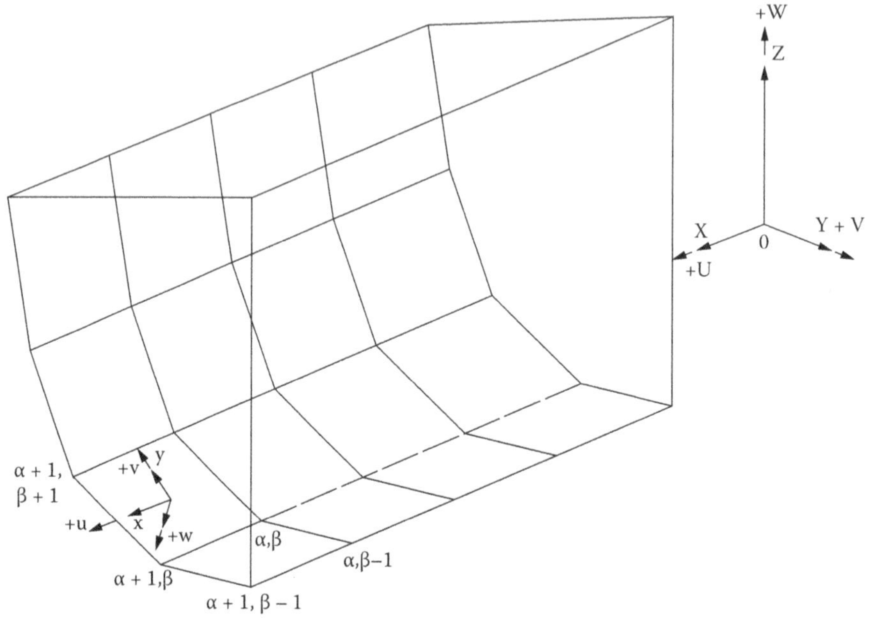

FIGURE 6.1
Shell breakdown.

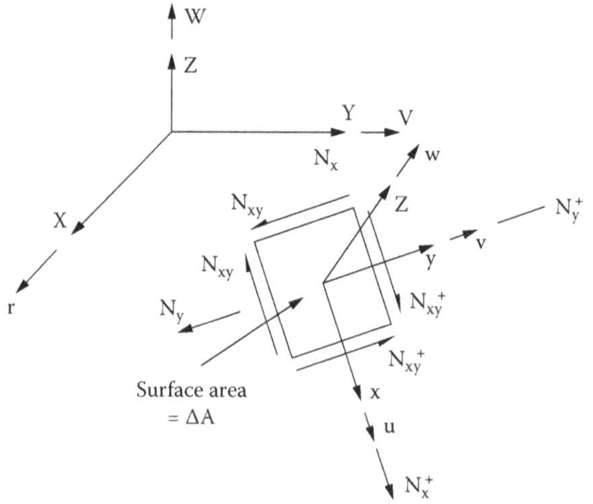

FIGURE 6.2
Local hull plate.

TABLE 6.1

Direction Cosines of Local Axes with Respect to Fixed Axes

	X	Y	Z
x	λ_1	μ_1	p_1
y	λ_2	μ_2	p_2
z	λ_3	μ_3	p_3

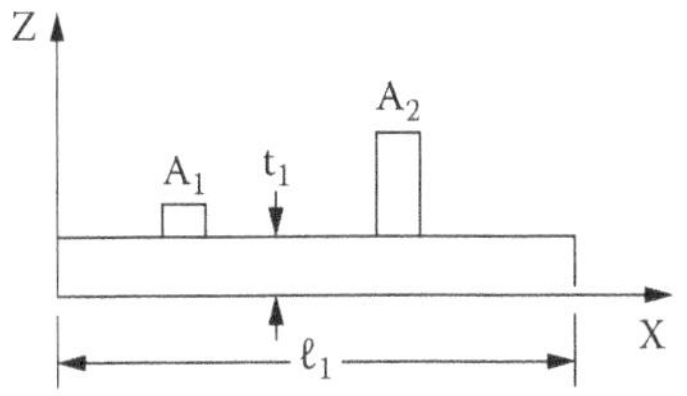

FIGURE 6.3
Sectional view of cross-stiffened plate.

and twisting. For the purpose of computing the energy of stretching of the stiffeners and plating, it will be assumed that the plating is fully effective. An equivalent thickness for tension and compression in the local x- and y-directions will be used and is computed as follows[2]:

$$h_x = \frac{A_x}{l_x}, \quad h_y = \frac{A_y}{l_y} \tag{6.2}$$

The parameters h_x and h_y are the equivalent thicknesses for stretching or compression in the local x- and y-directions, respectively. A_x and A_y are the total cross-sectional areas of the plating and stiffeners for load carrying in the x- and y-directions; l_x and l_y are the widths of the plating. For example, consider the plate shown in Figure 6.3. In this case, the nomenclature is as follows:

X, Y, Z: fixed rectangular axes; origin of these axes can be chosen at any convenient point; X-axis is chosen in longitudinal direction, Y-axis in athwart ship's direction, and Z-axis vertical

α, β: a set of numbers used to locate hull shell elements; α and β are actually two numbers by means of which the position of a point on a surface can be expressed

x, y, z: local coordinate axes associated with each element; x, y are in plane of element, and z is perpendicular to plane of element

λ_1, μ_1, p_1: direction cosines of local x-axis of hull plate element with respect to X-, Y-, and Z-axes, respectively

λ_2, μ_2, p_2: direction cosines of local y-axis of hull plate element with respect to X-, Y-, and Z-axes, respectively

λ_3, μ_3, p_3: direction cosines of local z-axis of hull plate element with respect to X-, Y-, and Z-axes, respectively

$U_{\alpha\beta}$, $V_{\alpha\beta}$, $W_{\alpha\beta}$: displacements of point α, β with respect to directions X, Y, and Z, respectively

T: total kinetic energy

P: total potential energy

Q, R, S: generalized forces arising from X-, Y-, and Z-components of applied force, respectively

N_x, N_y: stretching or compressive force (per unit length) in local x-, y-direction

N_{xy}: shear force per unit length

ΔA: surface area of plate element (equal to length × width); $\Delta A_{\alpha\beta}$ is surface area of hull plate element

u, v, w: displacements in local x-, y-, and z-directions, respectively, of a point on an elastic element

h_x, h_y: equivalent thickness for stretching in local x- and y-directions, respectively; these thicknesses are calculated by assuming that cross-sectional area of actual plate, including stiffeners, is equal to cross-sectional area of equivalent-thickness plate

A_x, A_y: total cross-sectional area of a cross-stiffened plate for load carrying in x- and y-directions, respectively

l_x, l_y: width of stiffened plate perpendicular to local x- and y-directions, respectively

ε_y, ε_y, γ_{xy}: strains in plating relative to local x- and y-axes

t: thickness of plating in a cross-stiffened plate

E: modulus of elasticity

v: Poisson's ratio

G: shear modulus

A: cross-sectional area of a bar element to be distinguished from ΔA, which is surface area of a cross-stiffened plate element; also to be distinguished from A_x and A_y, which are cross-sectional areas of a cross-stiffened plate

x: length of plate or bar element in local x direction

y: length of plate or bar element in local y direction

D_x, D_y, D_1, D_{xy}: orthotropic constants for bending

ω: frequency

k: stretching stiffness coefficients

$\bar{k}$: bending stiffness coefficients

c_o: sound velocity in water

$$A_v = A_1 + A_2 + t_1 l$$
$$l_v = l_1 \tag{6.3}$$
$$h_v = \frac{A_1 + A_2 + t_1 l_1}{l_1} = \frac{A_1 + A_2}{l_1} + t_1$$

Thus, the equivalent thickness for stretching or compression in a given direction is equal to the actual thickness plus a factor depending on the load-carrying area of the stiffeners and the width of the plate. For purposes of computing the energy of shearing, the thickness of the plating alone will be used.

The membrane potential energy can then be written as follows (relative to the local x-, y-, z-system; Figure 4.2; reference 3):

$$E_x = \int_A \frac{1}{2}(N_x \in_x + N_y \in_y + N_{xy}\gamma_{xy})dA$$
$$\in_x = \frac{\partial u}{\partial x}, \quad \in_y = \frac{\partial v}{\partial y}, \quad \gamma_{xy} = \frac{\partial u}{\partial y} + \frac{\partial v}{\partial x} \tag{6.4}$$
$$\in_x = \frac{1}{E}\left[\frac{N_x}{h_x} - \gamma\frac{N_v}{h_y}\right], \quad \in_v = \frac{1}{E}\left[\frac{N_v}{h_v} - v\frac{N_x}{h_x}\right],$$
$$\gamma_{xy} = \frac{1}{Gl}N_{xy}$$

The total potential energy of stretching and shearing can then be written as follows:

$$\Delta E_x = \left\{\frac{1}{2}\frac{E}{1-\nu^2}\left[h_x\left(\frac{\partial u}{\partial x}\right)^2 + h_v\left(\frac{\partial v}{\partial y}\right)^2 \right.\right.$$
$$\left.\left. + v(h_x + h_y)\left(\frac{\partial u}{\partial x}\right)\left(\frac{\partial v}{\partial y}\right)\right] + \frac{Gt}{2}\left(\frac{\partial v}{\partial x} + \frac{\partial u}{\partial y}\right)^2\right\}\Delta A \tag{6.5}$$

Expressing the derivatives in finite-difference form, the total potential energy of stretching and shearing relative to the fixed axes can be written as shown in Section 6.7.

6.2.1 Bending and Twisting Energy

The potential energy of bending and twisting of a cross-stiffened plate element can be derived from the concept of an equivalent orthotropic plate. The energy is

$$E_B = \int_A \left[\frac{1}{2} D_x \left(\frac{\partial^2 \omega}{\partial x^2} \right)^2 + D_v \left(\frac{\partial^2 \omega}{\partial y^2} \right)^2 + 2D_1 \frac{\partial^2 w}{\partial x^2} \frac{\partial^2 w}{\partial y^2} + 4D_{xy} \left(\frac{\partial^2 w}{\partial x \partial y} \right)^2 \right] dA \tag{6.6}$$

Using finite-difference approximations and transferring to the fixed axes (Equation 6.12), the potential energy of bending and twisting of a plate element can be written as shown in Section 6.8.

6.3 Equations of Motion and Basic Matrix

Using the three displacements, U, V, and W, as generalized coordinates, the Lagrange equations of motion of the system of plates making up the built-up shell are

$$\frac{d}{dt}\left(\frac{\partial T}{\partial U} \right) + \frac{\partial P}{\partial U} = Q_U \tag{6.7}$$

and similarly for V and W.

In Equation 6.7, T is the kinetic energy, P is the potential energy, and Q is the tile generalized force. The kinetic energy for any element can be written as

$$T = \tfrac{1}{2} m(\dot{U}^2 + \dot{V}^2 + \dot{W}^2) \tag{6.8}$$

where m is the mass of the element.

Substituting the kinetic energy and performing the indicated operations results in a set of simultaneous linear differential equations in tiles U, V, and W. For the free-vibration case, the generalized forces are zero and the equations are homogeneous. The problem reduces to an eigenvalue problem for determination of the eigenvalues and eigenvectors of the structure. The eigenvalue matrix is very sparsely populated since the equation of motion written for point involves only the points immediately surrounding α, β.

6.4 Effect of Fluid

It will be assumed that each rectangular plate element in contact with the water has uniform velocity over its surface. It therefore can be considered a rectangular piston located in a baffle, which has the shape of the hull. The total pressure on each plate element will be the pressure due to the component of its own motion perpendicular to the water plus the pressure on it due to all the other elements vibrating in the water. As a first approximation, the pressure on each plate element due to its own motion alone can be used as explained in Chapter 5.

6.5 Modal Impedances

If the geometry of the structure can be described analytically, it may sometimes be possible to determine the pressure for a given mode shape. The forced vibrations problem can then be handled mode by mode and the entire solution built up of a superposition of modes. Methods for determining these modal impedances for infinite cylinders are given by Junger[4] and for spheroids by Junger and Kleinschmidt[5] and Chertock.[6]

6.6 Calculation of the Green's Function

The basic Lagrange equations for the displacements $U_{\alpha\beta}$, $V_{\alpha\beta}$, and $W_{\alpha\beta}$ are

$$P = E_S + E_B \quad T = \frac{1}{2}\sum_{\alpha}\sum_{\beta} m_{\alpha\beta}(U_{\alpha\beta}^2 + V_{\alpha\beta}^2 + W_{\alpha\beta}^2)$$

$$\frac{d}{dt}\left(\frac{\partial T}{\partial U_{\alpha\beta}}\right) + \frac{\partial \rho}{\partial U_{\alpha\beta}} = Q_U; \quad \frac{d}{dt}\left(\frac{\partial T}{\partial V_{\alpha\beta}}\right) + \frac{\partial P}{\partial V_{\alpha\beta}} = Q_V; \quad \frac{d}{dt}\left(\frac{\partial T}{\partial W_{\alpha\beta}}\right) + \frac{\partial P}{\partial W_{\alpha\beta}} = Q_W \tag{6.9}$$

where T is the kinetic energy and P is the potential energy ($E_S + E_B$). The Q_U, Q_V, and Q_W are the vector components of the harmonic force $F_o e^{iwt}$.

If water loading is included via the piston approximation, then the equations have the following added terms on the left side of each equation, respectively.

$$A\rho_o c_o(\theta + i\chi)\lambda_3 \dot{U},\ A\rho_o c_o(\theta + i\chi)\mu_3 \dot{V},\ A\rho_o c_o(\theta + i\chi)\rho_3 \dot{W} \tag{6.10}$$

where $\dot{U} = i\omega U e^{i\omega t}$, $\dot{V} = i\omega V e^{i\omega t}$, $\dot{W} = i\omega W e^{i\omega t}$ and A = area of the plate element.

This will give a set of complex simultaneous algebraic equations for the displacements $U_{\alpha\beta}$, $V_{\alpha\beta}$, and $W_{\alpha\beta}$. Once the displacements are computed, the accelerations are just ω^2 times the displacements and the far field can be computed by the methods outlined in Chapter 13.

6.7 Potential Energy of Stretching and Shearing

The total potential energy of stretching and shearing relative to the fixed axes can be written in finite-difference form as

$$E_B = \sum_{\alpha\beta} \frac{\Delta A_{\alpha\beta}}{2} \frac{E_{\alpha\beta}}{1-\nu^2} \left\{ (h_r)_{\alpha\beta} \left(\frac{\begin{array}{c}[(\lambda_1)_{\alpha\beta} U_{\alpha+1\beta} + (\mu_1)_{\alpha\beta} V_{\alpha+1\beta} + (p_1)_{\alpha\beta} W_{\alpha+1\beta}] \\ -[(\lambda_1)_{\alpha\beta} U_{\alpha\beta} + (\mu_1)_{\alpha\beta} V_{\alpha\beta} + (p_1)_{\alpha\beta} W_{\alpha\beta}]\end{array}}{\Delta x_{\alpha\beta}} \right)^2 \right.$$

$$+ (h_y)_{\alpha\beta} \left(\frac{\begin{array}{c}[(\lambda_2)_{\alpha\beta} U_{\alpha\beta+1} + (\mu_2)_{\alpha\beta} V_{\alpha\beta+1} + (p_2)_{\alpha\beta} W_{\alpha\beta+1}] \\ -[(\lambda_2)_{\alpha\beta} U_{\alpha\beta} + (\mu_2)_{\alpha\beta} V_{\alpha\beta} + (p_2)_{\alpha\beta} W_{\alpha\beta}]\end{array}}{\Delta y_{\alpha\beta}} \right)^2$$

$$+ \nu[(h_z)_{\alpha\beta} + (h_y)_{\alpha\beta}] \left(\frac{\begin{array}{c}[(\lambda_1)_{\alpha\beta} U_{\alpha+1\beta} + (\mu_1)_{\alpha\beta} V_{\alpha+1\beta} + (p_1)_{\alpha\beta} W_{\alpha+1\beta}] \\ -[(\lambda_1)_{\alpha\beta} U_{\alpha\beta} + (\mu_1)_{\alpha\beta} V_{\alpha\beta} + (p_1)_{\alpha\beta} W_{\alpha\beta}]\end{array}}{\Delta x_{\alpha\beta}} \right)$$

$$\left. \times \left(\frac{\begin{array}{c}[(\lambda_2)_{\alpha\beta} U_{\alpha\beta+1} + (\mu_2)_{\alpha\beta} V_{\alpha\beta+1} + (p_2)_{\alpha\beta} W_{\alpha\beta+1}] \\ -[(\lambda_2)_{\alpha\beta} U_{\alpha\beta} + (\mu_2)_{\alpha\beta} V_{\alpha\beta} + (p_2)_{\alpha\beta} W_{\alpha\beta}]\end{array}}{\Delta y_{\alpha\beta}} \right) \right\}$$

$$+\frac{\Delta A_{\alpha\beta}}{2} G_{\alpha\beta} t_{\alpha\beta} \left[\left(\frac{[(\lambda_2)_{\alpha\beta} U_{\alpha+1\beta} + (\mu_2)_{\alpha\beta} V_{\alpha+1\beta} + (p_1)_{\alpha\beta} W_{\alpha+1\beta}] - [(\lambda_2)_{\alpha\beta} U_{\alpha\beta} + (\mu_2)_{\alpha\beta} V_{\alpha\beta} + (p_2)_{\alpha\beta} W_{\alpha\beta}]}{\Delta x_{\alpha\beta}} \right) \right.$$

$$\left. + \left(\frac{[(\lambda_1)_{\alpha\beta} U_{\alpha\beta+1} + (\mu_1)_{\alpha\beta} V_{\alpha\beta+1} + (p_1)_{\alpha\beta} W_{\alpha\beta+1}] - [(\lambda_1)_{\alpha\beta} U_{\alpha\beta} + (\mu_1)_{\alpha\beta} V_{\alpha\beta} + (p_1)_{\alpha\beta} W_{\alpha\beta}]}{\Delta y_{\alpha\beta}} \right) \right]^2 \right\} \tag{6.11}$$

6.8 Potential Energy of Bending and Twisting

Using finite-difference approximations and transforming to the fixed axes, the potential energy of bending and twisting of a plate element is

$$E_B = \frac{1}{2} \sum_{\alpha\beta} \left\{ [D_x]_{\alpha\beta} \frac{\begin{aligned}&[(\lambda_\delta)_{\alpha\beta} U_{\alpha+1\beta} + (\mu_\delta)_{\alpha\beta} V_{\alpha+1\beta} + (p_\delta)_{\alpha\beta} W_{\alpha+1\beta} \\ &-2(\lambda_\delta)_{\alpha\beta} U_{\alpha\beta} - 2(\mu_\delta)_{\alpha\beta} V_{\alpha\beta} - 2(p_\delta)_{\alpha\beta} W_{\alpha\beta} + (\lambda_\delta)_{\alpha\beta} U_{\alpha-1\beta} \\ &+(\mu_\delta)_{\alpha\beta} V_{\alpha-1\beta} + (p_\delta)_{\alpha\beta} W_{\alpha-1\beta}]^2\end{aligned}}{(\Delta x_{\alpha\beta})^4} \right.$$

$$+ [D_y]_{\alpha\beta} \frac{\begin{aligned}&[(\lambda_\delta)_{\alpha\beta} U_{\alpha\beta+1} + (\mu_\delta)_{\alpha\beta} V_{\alpha\beta+1} + (p_\delta)_{\alpha\beta} W_{\alpha\beta+1} - 2(\lambda_\delta)_{\alpha\beta} U_{\alpha\beta} \\ &-2(\mu_\delta)_{\alpha\beta} V_{\alpha\beta} - 2(p_\delta)_{\alpha\beta} W_{\alpha\beta} + (\lambda_\delta)_{\alpha\beta} U_{\alpha\beta-1} + (\mu_\delta)_{\alpha\beta} V_{\alpha\beta-1} \\ &+(p_\delta)_{\alpha\beta} W_{\alpha\beta-1}]^2\end{aligned}}{(\Delta y_{\alpha\beta})^4}$$

$$+ 4[D_{zy}]_{\alpha\beta} \frac{\begin{aligned}&[(\lambda_\delta)_{\alpha\beta} U_{\alpha\beta+1} + (\mu_\delta)_{\alpha\beta} V_{\alpha\beta+1} + (p_\delta)_{\alpha\beta} W_{\alpha\beta+1} - (\lambda_\delta)_{\alpha\beta} U_{\alpha\beta} \\ &-(\mu_3)_{\alpha\beta} V_{\alpha\beta} - (p_\delta)_{\alpha\beta} W_{\alpha\beta} - (\lambda_\delta)_{\alpha\beta} U_{\alpha-1\beta+1} - (\mu_\delta)_{\alpha\beta} V_{\alpha-1\beta+1} \\ &-(p_\delta)_{\alpha\beta} W_{\alpha-1\beta+1} + (\lambda_\delta)_{\alpha\beta} U_{\alpha-1\beta} + (\mu_\delta)_{\alpha\beta} V_{\alpha-1\beta} + (p_\delta)_{\alpha\beta} W_{\alpha-1\beta}]^2\end{aligned}}{(\Delta x_{\alpha\beta} \Delta y_{\alpha\beta})^2}$$

$$+ 2[D_1]_{\alpha\beta} \frac{\begin{aligned}&[(\lambda_\delta)_{\alpha\beta} U_{\alpha+1\beta} + (\mu_\delta)_{\alpha\beta} V_{\alpha+1\beta} + (p_\delta)_{\alpha\beta} W_{\alpha+1\beta} - 2(\lambda_\delta)_{\alpha\beta} U_{\alpha\beta} \\ &-2(\mu_\delta)_{\alpha\beta} V_{\alpha\beta} - 2(p_\delta)_{\alpha\beta} W_{\alpha\beta} + (\lambda_\delta)_{\alpha\beta} U_{\alpha-1\beta} \\ &+(\mu_\delta)_{\alpha\beta} V_{\alpha-1\beta} + (p_\delta)_{\alpha\beta} W_{\alpha-1\beta}]\end{aligned}}{(\Delta x_{\alpha\beta})^2}$$

$$\left.\times\frac{\begin{array}{l}[\lambda_\delta]_{\alpha\beta}U_{\alpha\beta+1}+(\mu_\delta)_{\alpha\beta}V_{\alpha\beta+1}+(p_\delta)_{\alpha\beta}W_{\alpha\beta+1}-(2\lambda_\delta)_{\alpha\beta}U_{\alpha\beta}-2(\mu_\delta)_{\alpha\beta}V_{\alpha\beta}\\ -2(p_\delta)_{\alpha\beta}W_{\alpha\beta}+(\lambda_\delta)_{\alpha\beta}U_{\alpha\beta-1}+(\mu_\delta)_{\alpha\beta}V_{\alpha\beta-1}+(p_\delta)_{\alpha\beta}W_{\alpha\beta-1}]\end{array}}{(\Delta y_{\alpha\beta})^2}\right\}\Delta A_{\alpha\beta} \tag{6.12}$$

6.9 Stiffness Coefficients

The stiffness coefficients for an individual element are determined by differentiating the total potential energy with respect to the displacements U, V, and W of that element. Thus,

$$\begin{aligned}\frac{\partial P}{\partial U_{\alpha\beta}}=\frac{\partial(E_S+E_B)}{\partial U_{\alpha\beta}}\\
&=\Big\{(\bar{k}_{\alpha\beta,\,\alpha-2\beta})^U_U U_{\alpha-2\beta}+[(\bar{k}_{\alpha\beta,\,\alpha-1\beta})^U_U+(k_{\alpha\beta,\,\alpha-1\beta})^U_U]U_{\alpha-1\beta}+(\bar{k}_{\alpha\beta,\,\alpha+2\beta})^U_U U_{\alpha+2\beta}\\
&\quad+[(\bar{k}_{\alpha\beta,\,\alpha\beta})^U_U+(k_{\alpha\beta,\,\alpha\beta})^U_U]U_{\alpha\beta}+[(\bar{k}_{\alpha\beta,\,\alpha+1\beta})^U_U+(k_{\alpha\beta,\,\alpha+1\beta})^U_U]U_{\alpha+1\beta}\\
&\quad+(\bar{k}_{\alpha\beta,\,\alpha\beta-2})^U_U U_{\alpha\beta-2}+[(\bar{k}_{\alpha\beta,\,\alpha\beta-1})^U_U+(k_{\alpha\beta,\,\alpha\beta-1})^U_U]U_{\alpha\beta-1}\\
&\quad+(\bar{k}_{\alpha\beta,\,\alpha\beta+2})^U_U U_{\alpha\beta+2}+[(\bar{k}_{\alpha\beta,\,\alpha\beta+1})^U_U+(k_{\alpha\beta,\,\alpha\beta+1})^U_U]U_{\alpha\beta+1}\\
&\quad+(\bar{k}_{\alpha\beta,\,\alpha-1\;\beta-1})^U_U U_{\alpha-1\;\beta-1}+[(\bar{k}_{\alpha\beta,\,\alpha-1\;\beta+1})^U_U+(k_{\alpha\beta,\,\alpha-1\;\beta+1})^U_U]U_{\alpha-1\;\beta+1}\\
&\quad+(\bar{k}_{\alpha\beta,\,\alpha+1\;\beta+1})^U_U U_{\alpha+1\;\beta+1}+(\bar{k}_{\alpha\beta,\,\alpha+1\;\beta-1})^U_U+(k_{\alpha\beta,\,\alpha+1\;\beta-1})^U_U U_{\alpha+1\;\beta-1}\Big\}\\
&\quad+\Big\{(\bar{k}_{\alpha\beta,\,\alpha-2\beta})^U_V V_{\alpha-2\beta}+\dots+[(\bar{k}_{\alpha\beta,\,\alpha+1\;\beta-1})^U_V+(k_{\alpha\beta,\,\alpha+1\;\beta-1})^U_V]V_{\alpha+1\;\beta-1}\Big\}\\
&\quad+\Big\{(\bar{k}_{\alpha\beta,\,\alpha-2\beta})^U_W W_{\alpha-2\beta}+\dots+[(\bar{k}_{\alpha\beta,\,\alpha+1\;\beta-1})^U_W+(k_{\alpha\beta,\,\alpha+1\;\beta-1})^U_W]W_{\alpha+1\;\beta-1}\Big\}\end{aligned} \tag{6.13}$$

$\partial P/\partial V_{\alpha\beta}$ and $\partial P/\partial W_{\alpha\beta}$ are obtained from $\partial P/\partial U_{\alpha\beta}$ by changing the superscript U in $(k)^U_U$ and $(\bar{k})^U_U$ to V and W, respectively.

The values of the k and $\bar{k}$ are

$$
\begin{aligned}
(k_{\alpha\beta,\,\alpha-1\beta})_U^U &= \Delta A_{\alpha-1\beta}\frac{E_{\alpha-1\beta}}{1-\nu^2}(h_x)_{\alpha-1\beta}(A')_U^U(A)_U^U \\
&\quad+\frac{\Delta A_{\alpha-1\beta}}{2}\frac{E_{\alpha-1\beta}}{1-\nu^2}\nu[(h_x)_{\alpha-1\beta}+(h_y)_{\alpha-1\beta}](B')_U^U(B)_U^U \\
&\quad+\Delta A_{\alpha-1\beta}G_{\alpha-1\beta}t_{\alpha-1\beta}(C')_U^U(C)_U^U \\
(k_{\alpha\beta,\alpha\beta})_U^U &= \frac{\Delta A_{\alpha\beta}E_{\alpha\beta}}{1-\nu^2}\Bigg[(h_x)_{\alpha\beta}(D')_U^U(D)_U^U+(h_y)_{\alpha\beta}(E')_U^U(E)_U^U \\
&\qquad+\frac{\nu[(h_x)_{\alpha\beta}+(h_y)_{\alpha\beta}}{2}\left((F')_U^U(F)_U^U-(G')_U^U(G)_U^U\right)\Bigg] \\
&\quad+\Delta A_{\alpha\beta}G_{\alpha\beta}t_{\alpha\beta}(H')_U^U(H)_U^U+\Delta A_{\alpha-1\beta}\frac{E_{\alpha-1\beta}}{1-\nu^2}(h_x)_{\alpha-1\beta}(-A')_U^U(A)_U^U \\
&\quad+\Delta A_{\alpha-1\beta}G_{\alpha-1\beta}t_{\alpha-1\beta}(C'')_U^U(C)_U^U+\frac{\Delta A_{\alpha\beta-1}E_{\alpha\beta-1}}{1-\nu^2}(h_y)_{\alpha\beta-1}(I')_U^U(I)_U^U \\
&\quad+\Delta A_{\alpha\beta-1}G_{\alpha\beta-1}t_{\alpha\beta-1}(J')_U^U(J)_U^U \\
(k_{\alpha\beta,\,\alpha+1\beta})_U^U &= \frac{\Delta A_{\alpha\beta}E_{\alpha\beta}}{1-\nu^2}\left[(h_x)_{\alpha\beta}(-D')_U^U(D)_U^U+\frac{\nu[(h_x)_{\alpha\beta}+(h_x)_{\alpha\beta}}{2}(-D')_U^U(F)_U^U\right] \\
&\quad+\Delta A_{\alpha\beta}G_{\alpha\beta}t_{\alpha\beta}(H'')_U^U(H)_U^U \\
(k_{\alpha\beta,\,\alpha\beta-1})_U^U &= \frac{\Delta A_{\alpha\beta-1}E_{\alpha\beta-1}}{1-\nu^2}(h_y)_{\alpha\beta-1}(-I')_U^U(I)_U^U-\frac{\Delta A_{\alpha\beta-1}E_{\alpha\beta-1}}{2(1-\nu^2)}\nu[(h_x)_{\alpha\beta-1} \\
&\quad+(h_y)_{\alpha\beta-1}](K')_U^U(K)_U^U+\Delta A_{\alpha\beta-1}G_{\alpha\beta-1}t_{\alpha\beta-1}(J'')_U^U(J)_U^U \\
(k_{\alpha\beta,\,\alpha\beta+1})_U^U &= \frac{\Delta A_{\alpha\beta}E_{\alpha\beta}}{1-\nu^2}(h_y)_{\alpha\beta}(-E')_U^U(E)_U^U+\frac{\Delta A_{\alpha\beta}}{2}\frac{E_{\alpha\beta}}{1-\nu^2} \\
&\quad\times\nu[(h_x)_{\alpha\beta}+(h_y)_{\alpha\beta}](-E')_U^U(G)_U^U+\Delta A_{\alpha\beta}G_{\alpha\beta}t_{\alpha\beta}(H''')_U^U(H)_U^U \\
(k_{\alpha\beta,\,\alpha-1\,\beta+1})_U^U &= \frac{\Delta A_{\alpha-1\beta}E_{\alpha-1\beta}}{2(1-\nu^2)}\nu[(h_x)_{\alpha-1\beta}+(h_y)_{\alpha-1\beta}](-B')_U^U(B)_U^U \\
&\quad+\Delta A_{\alpha-1\beta}G_{\alpha-1\beta}t_{\alpha-1\beta}(C''')_U^U(C)_U^U \\
(k_{\alpha\beta,\,\alpha+1\,\beta-1})_U^U &= \frac{\Delta A_{\alpha\beta-1}E_{\alpha\beta-1}}{2(1-\nu^2)}\nu[(h_x)_{\alpha\beta-1}+(h_y)_{\alpha\beta-1}](-K')_U^U(K)_U^U \\
&\quad+\Delta A_{\alpha\beta-1}G_{\alpha\beta-1}t_{\alpha\beta-1}(J''')_U^U(J)_U^U
\end{aligned}
\tag{6.14}
$$

where

$(A)_U^U = (\lambda_1)_{\alpha-1\beta}/\Delta x_{\alpha-1\beta}$	$(A')_U^U = -(\lambda_1)_{\alpha-1\beta}/\Delta x_{\alpha-1\beta}$	$(C'')_U^U = (\lambda_2)_{\alpha-1\beta}/\Delta x_{\alpha-1\beta}$
$(B)_U^U = (\lambda_1)_{\alpha-1\beta}/\Delta x_{\alpha-1\beta}$	$(B')_U^U = -(\lambda_2)_{\alpha-1\beta}/\Delta y_{\alpha-1\beta}$	$(C''')_U^U = (\lambda_1)_{\alpha-1\beta}/\Delta y_{\alpha-1\beta}$
$(C)_U^U = (\lambda_2)_{\alpha-1\beta}/\Delta x_{\alpha-1\beta}$	$(C')_U^U = [-(\lambda_2)_{\alpha-1\beta}/\Delta x_{\alpha-1\beta}]$ $-[(\lambda_1)_{\alpha-1\beta}/\Delta y_{\alpha-1\beta}]$	$(H'')_U^U = (\lambda_2)_{\alpha\beta}/\Delta x_{\alpha\beta}$
$(D)_U^U = -(\lambda_1)_{\alpha\beta}/\Delta x_{\alpha\beta}$	$(D')_U^U = -(\lambda_1)_{\alpha\beta}/\Delta x_{\alpha\beta}$	$(H''')_U^U = (\lambda_1)_{\alpha\beta}/\Delta y_{\alpha\beta}$
$(E)_U^U = -(\lambda_2)_{\alpha\beta}/\Delta y_{\alpha\beta}$	$(E')_U^U = -(\lambda_2)_{\alpha\beta}/\Delta y_{\alpha\beta}$	$(J'')_U^U = [-(\lambda_2)_{\alpha\beta-1}/\Delta x_{\alpha\beta-1}]$ $-[(\lambda_1)_{\alpha\beta-1}/\Delta y_{\alpha\beta-1}]$
$(F)_U^U = -(\lambda_2)_{\alpha\beta}/\Delta y_{\alpha\beta}$	$(F')_U^U = -(\lambda_1)_{\alpha\beta}/\Delta x_{\alpha\beta}$	$(J''')_U^U = (\lambda_2)_{\alpha\beta-1}/\Delta x_{\alpha\beta-1}$
$(G)_U^U = -(\lambda_1)_{\alpha\beta}/\Delta x_{\alpha\beta}$	$(G')_U^U = (\lambda_2)_{\alpha\beta}/\Delta y_{\alpha\beta}$	
$(H)_U^U = [-(\lambda_2)_{\alpha\beta}/\Delta x_{\alpha\beta}]$ $-[(\lambda_1)_{\alpha\beta}/\Delta y_{\alpha\beta}]$	$(H')_U^U = [-(\lambda_2)_{\alpha\beta}/\Delta x_{\alpha\beta}]$ $-[(\lambda_1)_{\alpha\beta}/\Delta y_{\alpha\beta}]$	
$(I)_U^U = (\lambda_2)_{\alpha\beta-1}/\Delta y_{\alpha\beta-1}$	$(I')_U^U = (\lambda_2)_{\alpha\beta-1}/\Delta y_{\alpha\beta-1}$	
$(J)_U^U = (\lambda_1)_{\alpha\beta-1}/\Delta y_{\alpha\beta-1}$	$(J')_U^U = (\lambda_1)_{\alpha\beta-1}/\Delta y_{\alpha\beta-1}$	
$(K)_U^U = (\lambda_2)_{\alpha\beta-1}/\Delta y_{\alpha\beta-1}$	$(K')_U^U = -(\lambda_1)_{\alpha\beta-1}/\Delta x_{\alpha\beta-1}$	

$$(\bar{k}_{\alpha\beta,\,\alpha-2\beta})_U^U = [D_x]_{\alpha-1\beta}\,\Delta A_{a-1\beta}(4')(4)$$

$$\begin{aligned}(\bar{k}_{\alpha\beta,\,\alpha-1\beta})_U^U = {} & [D_x]_{\alpha\beta}\,\Delta A_{\alpha\beta}(1')(1) + 4[D_{xy}]_{\alpha\beta}\,\Delta A_{\alpha\beta}(3')(3) + [D_1]_{\alpha\beta}\,\Delta A_{\alpha\beta}(2')(2) \\ & + [D_x]_{\alpha-1\beta}\,\Delta A_{\alpha-1\beta}[-2(4')](4) \\ & + [D_1]_{\alpha-1\beta}\,\Delta A_{\alpha-1\beta}(4'')(4) + 4[D_{xy}]_{\alpha\beta-1}\,\Delta A_{\alpha\beta-1}(6')(6)\end{aligned}$$

$$\begin{aligned}(\bar{k}_{\alpha\beta,\,\alpha\beta})_U^U = {} & [D_x]_{\alpha\beta}\,\Delta A_{\alpha\beta}(1'')(1) + [D_y]_{\alpha\beta}\,\Delta A_{\alpha\beta}(2'')(2) \\ & + 4[D_{xy}]_{\alpha\beta}\,\Delta A_{\alpha\beta}(-3')(3) - 2[D_1]_{\alpha\beta}\,\Delta A_{\alpha\beta}(2')(2) \\ & + [D_1]_{\alpha\beta}\,\Delta A_{\alpha\beta}(1''')(1) + [D_x]_{\alpha-1\beta}\,\Delta A_{\alpha-1\beta}(4')(4) + [D_y]_{\alpha\beta-1}\,\Delta A_{\alpha\beta-1}(5')(5) \\ & + 4[D_{xy}]_{\alpha\beta-1}\,\Delta A_{\alpha\beta-1}(-6')(6)\end{aligned}$$

$$+[D_x]_{\alpha+1\beta}\Delta A_{\alpha+1\beta}(7')(7)+4[D_{xy}]_{\alpha+1\beta}\Delta A_{\alpha+1\beta}(8')(8)$$
$$+[D_y]_{\alpha\beta+1}\Delta A_{\alpha\beta+1}(9')(9)+4[D_{xy}]_{\alpha+1\ \beta-1}\Delta A_{\alpha+1\ \beta-1}(10')(10)$$
$$(\bar{k}_{\alpha\beta,\ \alpha+1\beta})_U^U=[D_x]_{\alpha\beta}\Delta A_{\alpha\beta}(1')(1)+[D_1]_{\alpha\beta}\Delta A_{\alpha\beta}(2')(2)+[D_x]_{\alpha+1\beta}\Delta A_{\alpha+1\beta}2(7')(7)$$
$$+4[D_{xy}]_{\alpha+1\beta}\Delta A_{\alpha+1\beta}(-8')(8)$$
$$+[D_1]_{\alpha+1\beta}\Delta A_{\alpha+1\beta}(7'')(7)+4[D_{xy}]_{\alpha+1\beta-1}\Delta A_{\alpha+1\ \beta-1}(-10')(10)$$
$$(\bar{k}_{\alpha\beta,\ \alpha+2\beta})_U^U=[D_x]_{\alpha+1\beta}\Delta A_{\alpha+1\beta}(7')(7)$$
$$(\bar{k}_{\alpha\beta,\ \alpha\beta-2})_U^U=[D_y]_{\alpha\beta-1}\Delta A_{\alpha\beta-1}(5')(5)$$
$$(\bar{k}_{\alpha\beta,\ \alpha\beta-1})_U^U=[D_y]_{\alpha\beta}\Delta A_{\alpha\beta}(2''')(2)+[D_1]_{\alpha\beta}\Delta A_{\alpha\beta}(2''')(1)$$
$$+[D_y]_{\alpha\beta-1}\Delta A_{\alpha\beta-1}2(-5')(5)$$
$$+4[D_{xy}]_{\alpha\beta-1}\Delta A_{\alpha\beta-1}(6')(6)+[D_1]_{\alpha\beta-1}\Delta A_{\alpha\beta-1}(5'')(5)$$
$$+4[D_{xy}]_{\alpha+1\ \beta-1}\Delta A_{\alpha+1\ \beta-1}(-10')(10)$$
$$(\bar{k}_{\alpha\beta,\ \alpha\beta+1})_U^U=[D_y]_{\alpha\beta}\Delta A_{\alpha\beta}(2''')(2)$$
$$+4[D_{xy}]_{\alpha\beta}\Delta A_{\alpha\beta}(3')(3)+[D_1]_{\alpha\beta}\Delta A_{\alpha\beta}(2''')(1)$$
$$+4[D_{xy}]_{\alpha+1\beta}\Delta A_{\alpha+1\beta}(-8')(8)+[D_y]_{\alpha\beta+1}\Delta A_{\alpha\beta+1}2(-9')(9)$$
$$+[D_1]_{\alpha\beta+1}\Delta A_{\alpha\beta+1}(9'')(9)$$
$$(\bar{k}_{\alpha\beta,\ \alpha\beta+2})_U^U=[D_y]_{\alpha\beta+1}\Delta A_{\alpha\beta+1}(9')(9)$$
$$(\bar{k}_{\alpha\beta,\ \alpha-1\ \beta-1})_U^U=[D_1]_{\alpha-1\beta}\Delta A_{\alpha-1\beta}(4''')(4)$$
$$+4[D_{xy}]_{\alpha\beta-1}\Delta A_{\alpha\beta-1}(-6')(6)+[D_1]_{\alpha\beta-1}\Delta A_{\alpha\beta-1}(5''')(5)$$
$$(\bar{k}_{\alpha\beta,\ \alpha-1\ \beta+1})_U^U=4[D_{xy}]_{\alpha\beta}\Delta A_{\alpha\beta}(-3')(3)$$
$$+[D_y]_{\alpha-1\beta}\Delta A_{\alpha-1\beta}(4''')(4)+[D_1]_{\alpha\beta+1}\Delta A_{\alpha\beta+1}(9''')(9)$$
$$(\bar{k}_{\alpha\beta,\ \alpha+1\ \beta-1})_U^U=[D_1]_{\alpha\beta-1}\Delta A_{\alpha\beta-1}(5''')(5)+[D_1]_{\alpha+1\beta}\Delta A_{\alpha+1\beta}(7''')(7)$$
$$+4[D_{xy}]_{\alpha+1\ \beta-1}\Delta A_{\alpha+1\ \beta-1}(10')(10)$$
$$(\bar{k}_{\alpha\beta,\ \alpha+1\ \beta-1})_U^U=4[D_{xy}]_{\alpha+1\beta}\Delta A_{\alpha+1\beta}(8')(8)+[D_1]_{\alpha+1\beta}\Delta A_{\alpha+1\beta}(7''')(7)$$
$$+[D_1]_{\alpha\beta+1}\Delta A_{\alpha\beta+1}(9''')(9)$$

(6.15)

where

$(1) = -2(\lambda_\delta)_{\alpha\beta}/\Delta x_{\alpha\beta}^2$	$(1') = (\lambda_\delta)_{\alpha\beta}/\Delta x_{\alpha\beta}^2$	$(1'') = -2(\lambda_\delta)_{\alpha\beta}/\Delta x_{\alpha\beta}^2$
$(2) = -2(\lambda_\delta)/_{\alpha\beta}\Delta y_{\alpha\beta}^2$	$(2') = (\lambda_\delta)_{\alpha\beta}/\Delta x_{\alpha\beta}^2$	$(2'') = -2(\lambda_\delta)_{\alpha\beta}/\Delta y_{\alpha\beta}^2$
$(3) = -(\lambda_2)_{\alpha\beta}/\Delta x_{\alpha\beta}\Delta y_{\alpha\beta}$	$(3') = (\lambda_\delta)_{\alpha\beta}/\Delta x_{\alpha\beta}\Delta y_{\alpha\beta}$	$(4'') = -2(\lambda_\delta)_{\alpha-1\beta}/\Delta y_{\alpha-1\beta}^2$
$(4) = (\lambda_\delta)_{\alpha-1\beta}/\Delta x_{\alpha-1\beta}^2$	$(4') = (\lambda_\delta)_{\alpha-1\beta}/\Delta x_{\alpha-1\beta}^2$	$(5'') = -2(\lambda_\delta)_{\alpha\beta-1}/\Delta x_{\alpha\beta-1}^2$
$(5) = (\lambda_\delta)_{\alpha\beta-1}/\Delta y_{\alpha\beta-1}^2$	$(5') = (\lambda_\delta)_{\alpha\beta-1}/\Delta y_{\alpha\beta-1}^2$	$(7'') = -2(\lambda_\delta)_{\alpha+1\beta}/\Delta y_{\alpha+1\beta}^2$
$(6) = (\lambda_\delta)_{\alpha\beta-1}/\Delta x_{\alpha\beta-1}\Delta y_{\alpha\beta-1}$	$(6') = -(\lambda_\delta)_{\alpha\beta-1}/\Delta x_{\alpha\beta-1}\Delta y_{\alpha\beta-1}$	$(9'') = -2(\lambda_\delta)_{\alpha\beta+1}/\Delta x_{\alpha\beta+1}^2$
$(7) = (\lambda_\delta)_{\alpha+1\beta}/\Delta x_{\alpha+1\beta}^2$	$(7') = (\lambda_\delta)_{\alpha+1\beta}/\Delta x_{\alpha+1\beta}^2$	$(1''') = -2(\lambda_\delta)_{\alpha\beta}/\Delta y_{\alpha\beta}^2$
$(8) = (\lambda_\delta)_{\alpha+1\beta}/\Delta x_{\alpha+1\beta}\Delta y_{\alpha+1\beta}$	$(8') = (\lambda_\delta)_{\alpha+1\beta}/\Delta x_{\alpha+1\beta}\Delta y_{\alpha+1\beta}$	$(2''') = (\lambda_\delta)_{\alpha\beta}/\Delta y_{\alpha\beta}^2$
$(9) = (\lambda_\delta)_{\alpha\beta+1}/\Delta y_{\alpha\beta+1}^2$	$(9') = (\lambda_\delta)_{\alpha\beta+1}/\Delta y_{\alpha\beta+1}^2$	$(4''') = (\lambda_\delta)_{\alpha-1\beta}/\Delta y_{\alpha-1\beta}^2$
$(10) = -(\lambda_\delta)_{\alpha+1\,\beta-1}/\Delta x_{\alpha+1\,\beta-1}\Delta y_{\alpha+1\,\beta-1}$	$(10') = -(\lambda_\delta)_{\alpha+1\,\beta-1}/\Delta x_{\alpha+1\,\beta-1}\Delta y_{\alpha+1\,\beta-1}$	$(5''') = (\lambda_\delta)_{\alpha\beta-1}/\Delta x_{\alpha\beta-1}^2$
		$(7''') = (\lambda_\delta)_{\alpha+1\beta}/\Delta y_{\alpha+1\beta}^2$
		$(9''') = (\lambda_\delta)_{\alpha\beta+1}/\Delta x_{\alpha\beta+1}^2$

The coefficients $(k)_V^U$, $(k)_W^U$ are obtained from $(k)_U^U$ by putting μ and p, respectively, in place of λ in the primed, double primed, and triple primed expressions contained in $(k)_U^U$. The coefficients $(k)_U^V$, $(k)_V^V$, $(k)_W^V$ are obtained from $(k)_U^U$, $(k)_V^U$, $(k)_W^U$ by putting μ in place of λ in the unprimed expressions contained in $(k)_U^U$. Similarly, the coefficients $(k)_U^W$, $(k)_V^W$, $(k)_W^W$ are obtained from $(k)_U^U$, $(k)_V^U$, $(k)_W^U$, respectively, by putting p in place of λ in the unprimed expressions. Thus, all the coefficients are obtained from $(k)_U^U$ by cyclic interchange of direction cosines. The same rule holds true for the $\bar{k}$ expressions.

It should be noted that the derivative of the potential energy with respect to the displacement at point $\alpha\beta$ is a function of the properties of the elements surrounding $\alpha\beta$ only. Physically speaking, we can imagine a mass point at αβ attached to an array of springs, the constants of which are determined by the elastic properties of the plates surrounding αβ.

6.10 Example Calculations

As an example of the calculations with this method, consider the cases shown in Figures 6.4–6.11.

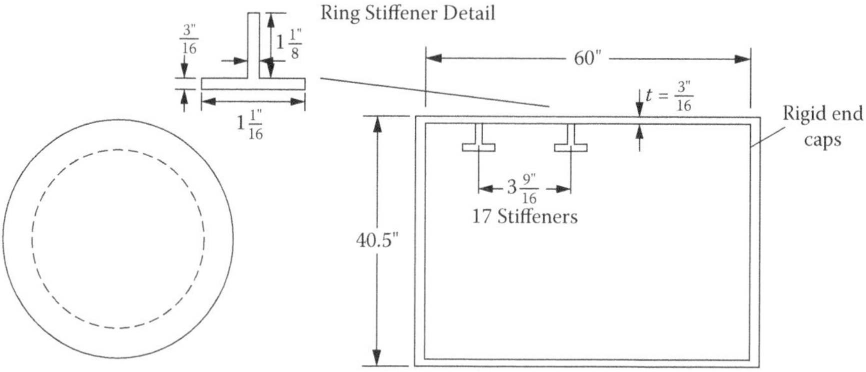

FIGURE 6.4
Sample calculation.

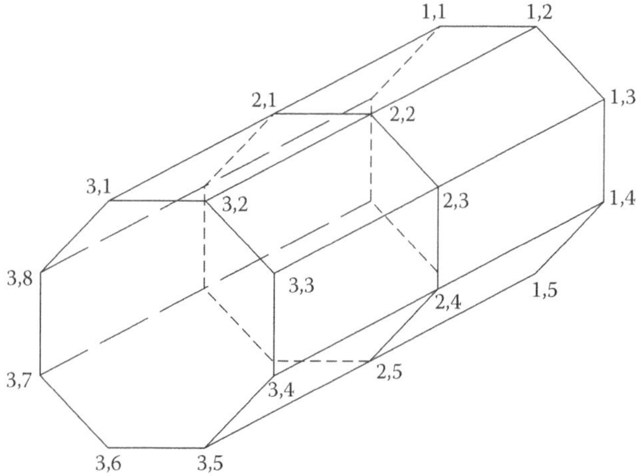

FIGURE 6.5
Sample calculation.

TABLE 6.2

$|K_{ij}| - \frac{Mw^2}{E}|I| = 0$ $M = 0.1045$ (lb-in. Sec Units) (Mass of Each Element) $E = 30 \times 10^6$ $|I|$ = Unit Matrix

	U_{21}	U_{22}	U_{23}	U_{24}	U_{25}	U_{26}	U_{27}	U_{28}	V_{21}	V_{22}	V_{23}	V_{24}	V_{25}	V_{26}	V_{27}	V_{28}	W_{21}	W_{22}	W_{23}	W_{24}	W_{25}	W_{26}	W_{27}	W_{28}
U_{21}	.4905	−.1361	0	0	0	0	0	−.1361	.1502	−.0780	0	0	0	0	0	−.0510	0	0	0	0	0	0	0	−.0510
U_{22}		.4905	−.1361	0	0	0	0	0	−.0721	.1062	−.0552	0	0	0	0	0	0	−.1062	.0552	0	0	0	0	0
U_{23}			.4905	−.1361	0	0	0	0	0	−.0510	0	0	0	0	0	0	0	.0510	−.1502	.0780	0	0	0	0
U_{24}				.4905	−.1361	0	0	0	0	0	0	−.1062	.0552	0	0	0	0	0	.0721	−.1062	.0552	0	0	0
U_{25}					.4905	−.1361	0	0	0	0	0	.0510	−.1502	.0780	0	0	0	0	0	.0510	0	0	0	0
U_{26}						.4905	−.1361	0	0	0	0	0	.0721	−.1062	.0552	0	0	0	0	0	0	.1062	−.0552	0
U_{27}							.4905	−.1361	0	0	0	0	0	.0510	0	0	0	0	0	0	0	−.0510	.1502	−.0780
U_{28}								.4905	−.0552	0	0	0	0	0	0	.1062	−.0552	0	0	0	0	0	−.0721	.1062
V_{21}									.9658	−.5930	0	0	0	0	0	−.2964	.2964	0	0	0	0	0	0	−.2964
V_{22}										.9276	−.2964	0	0	0	0	0	0	−.3346	.2964	0	0	0	0	0
V_{23}											.2964	0	0	0	0	0	0	.2964	−.2964	0	0	0	0	0
V_{24}												.3346	−.2964	0	0	0	0	0	0	.3346	−.2964	0	0	0
V_{25}													.9658	−.5930	0	0	0	0	0	−.2964	.2964	0	0	0
$\lvert K_{ij} \rvert =$ V_{26}				Symmetric Matrix										.9276	−.2964	0	0	0	0	0	0	−.3346	.2964	0
V_{27}															.2964	0	0	0	0	0	0	.2964	−.2964	0
V_{28}																.3346	−.2964	0	0	0	0	0	0	.3346
W_{21}																	.2964	0	0	0	0	0	0	−.2964
W_{22}																		.3346	−.2964	0	0	0	0	0
W_{23}																			.9658	−.5930	0	0	0	0
W_{24}																				.9276	−.2964	0	0	0
W_{25}																					.2964	0	0	0
W_{26}																						.3346	−.2964	0
W_{27}																							.9658	−.5930
W_{28}																								.9276

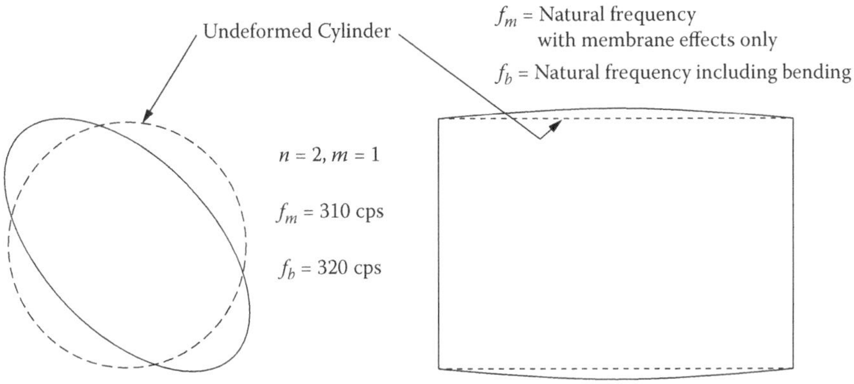

FIGURE 6.6
Sample calculation.

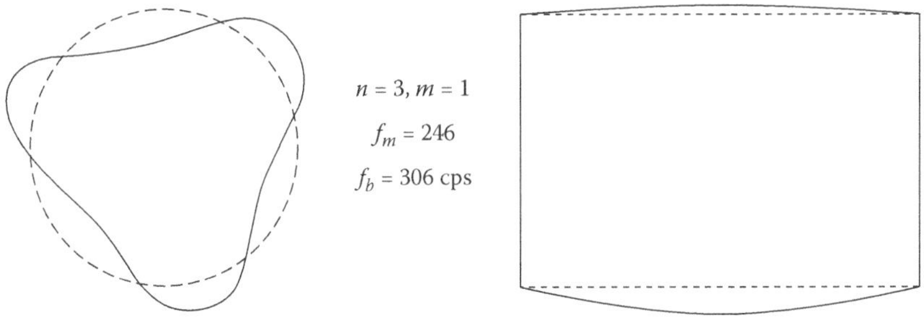

FIGURE 6.7
Sample calculation.

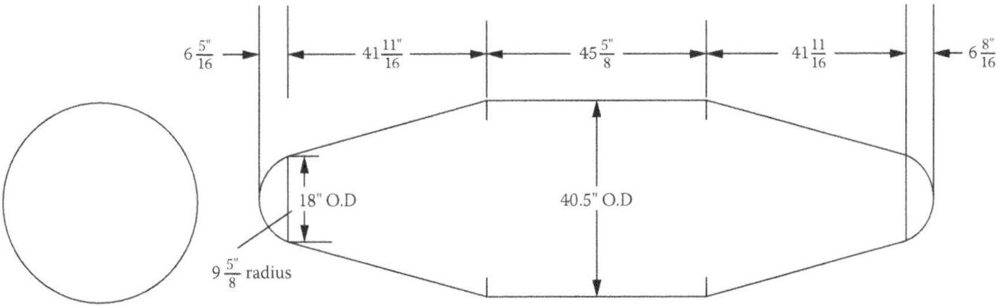

FIGURE 6.8
Sample calculation.

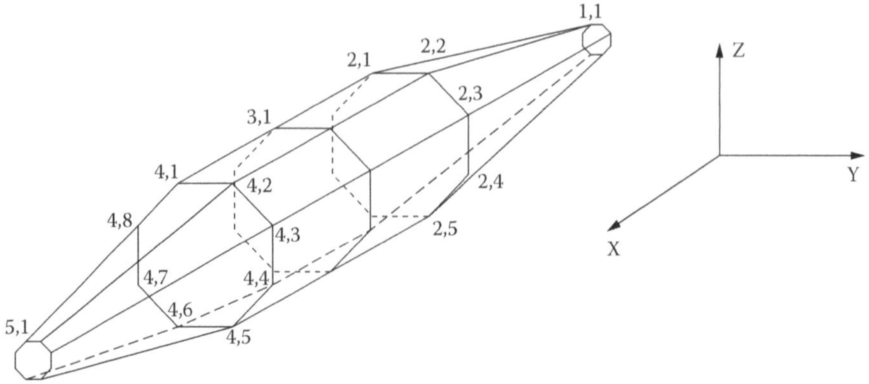

FIGURE 6.9
Sample calculation.

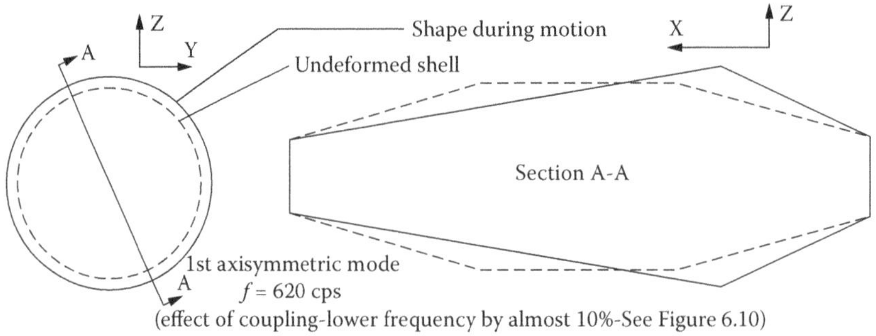

FIGURE 6.10
Sample calculation.

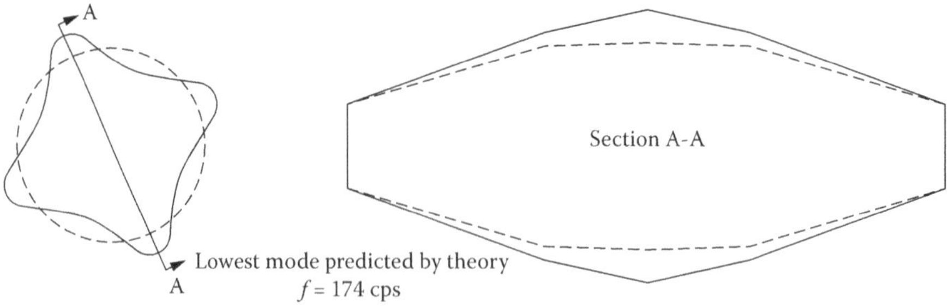

FIGURE 6.11
Sample calculation.

References

1. Greenspon, J. E. 1963. Theoretical developments in the vibrations of hulls. *Journal of Ship Research* 6 (4).
2. Kuhn, P. 1956. *Stresses in aircraft and shell structures*, 5–7. New York: McGraw-Hill Book Company, Inc.
3. Timoshenko, S. 1940. *Theory of plates and shells*, 1st ed., 7th impression, 303. New York: McGraw-Hill Book Company, Inc.
4. Junger, M. C. 1953. The physical interpretation of the expression for an outgoing wave in cylindrical coordinates. *Journal of the Acoustical Society of America* 25:40–47.
5. Junger, M. C., and Kleinschmidt, K. 1960. Acoustic mutual and self-radiation impedances for surface elements on a prolate spheroid, and associated far field sound pressures. Cambridge Acoustical Associates, Inc., done under contract with the Office of Naval Research.
6. Chertock, G. 1961. Sound radiation from prolate spheroids. *Journal of the Acoustical Society of America* 33:871–876.

7

Sound Patterns from Cylindrical Shells

7.1 Introduction

Junger,[1] Bleich and Baron,[2] and Greenspon[3] have presented the theory of the isotropic cylindrical shell vibrating in water. The purpose of these basic studies was to indicate the major parameters that affected the frequencies and the response of the shell. Ring-stiffened construction has been of interest for many years. However, only recently has attention been focused on orthotropic materials and sandwich type construction. The relative buckling strength of these types of structures is of primary consideration in their design, but the vibration and radiation characteristics is a parallel problem that must be investigated.

In practical situations, the cylindrical structure is subjected to many types of exciting forces. Most of these forces are not describable in any straightforward fashion and some are even random in nature. However, some of the exciting forces can usually be described statistically by correlation functions and power spectra. The purpose of this section is to give a unified presentation of the free and forced vibrations of stiffened and sandwich type cylindrical shells vibrating in water under general types of discrete and random forces.

7.2 Basic Equations

7.2.1 Types of Construction

In this section, the following types of construction will be considered[4]:

- Case 1: shells with rings and stringers (Figure 7.1)
- Case 2: sandwich type shells with various internal structures (Figure 7.2)

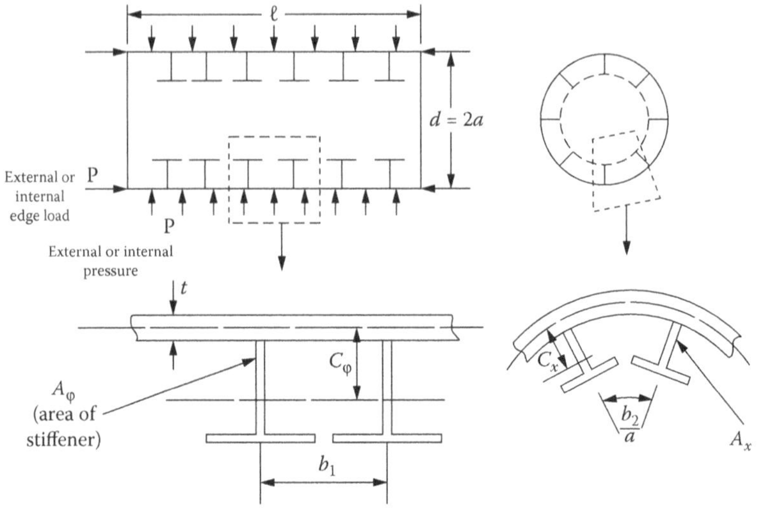

FIGURE 7.1
Shell with rings and stringers.

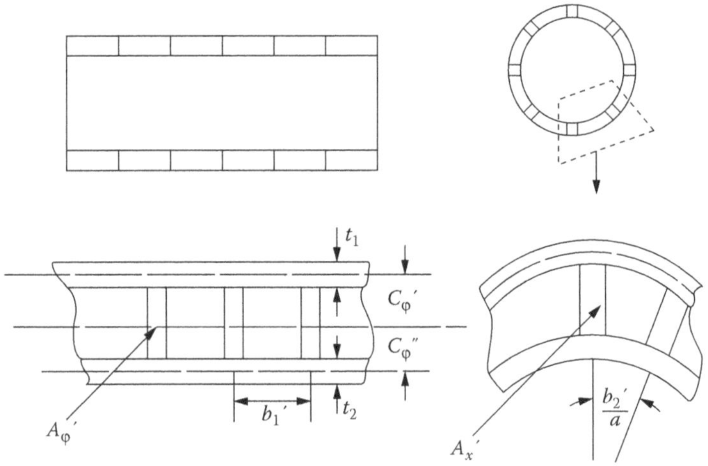

FIGURE 7.2
Sandwich shell.

- Case 3: solid shells made of high-strength orthotropic material (Figure 7.3)
- Case 4: isotropic shells (Figure 7.4)

The elastic relations between the stress or moment resultants and the displacements for the various cases are seen in Figure 7.5.

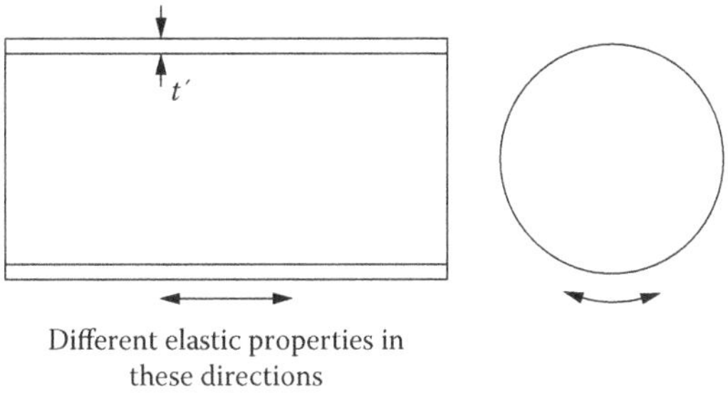

FIGURE 7.3
Shell of orthotropic material.

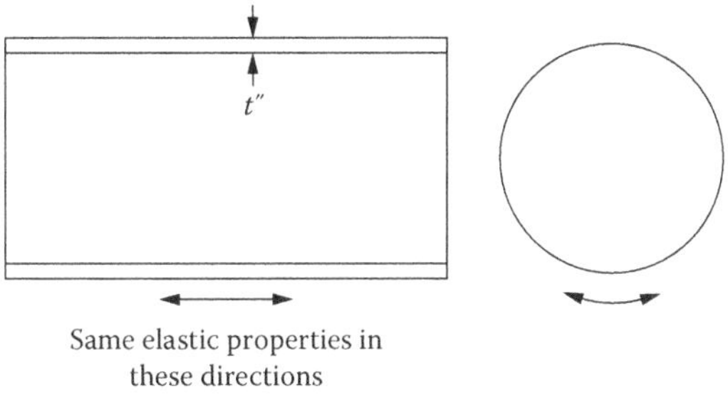

FIGURE 7.4
Isotropic shell.

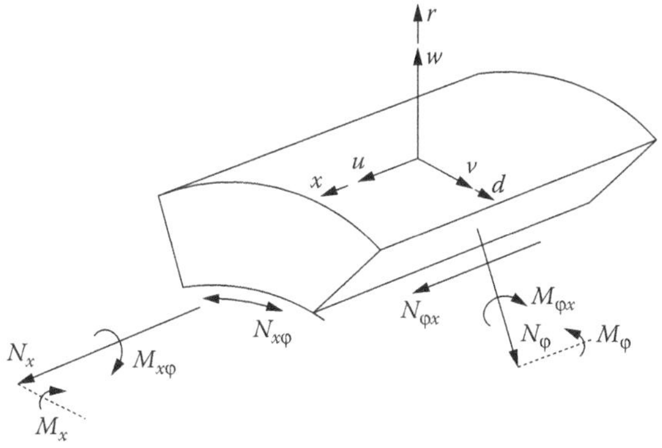

FIGURE 7.5
Force and moment resultants.

- Case 1: shells with rings and stringers

$$
\begin{aligned}
N_\varphi &= \frac{D_\varphi}{a}(v^{\cdot}+w)+\frac{D_\nu}{a}u'-\frac{S_\varphi}{a^2}w^{\cdot\cdot}\\
N_x &= \frac{D_x}{a}u'+\frac{D_\nu}{a}(v^{\cdot}+w)-\frac{S_x}{a^2}w''\\
N_{\varphi x} &= \frac{D_{\varphi x}}{a}(u^{\cdot}+v')+\frac{K_{\varphi x}}{a^3}w'^{\cdot}\\
N_{x\varphi} &= \frac{D_{x\varphi}}{a}(u^{\cdot}+v')\\
M_\varphi &= \frac{K_\varphi}{a^2}w^{\cdot\cdot}+\frac{K_\nu}{a^2}w''-\frac{S_\varphi}{a}(v^{\cdot}+w)\\
M_x &= \frac{K_x}{a^2}w''+\frac{K_\nu}{a^2}w''-\frac{S_x}{a}u'\\
M_{\varphi x} &= \frac{K_{\varphi x}}{a^2}w'^{\cdot}\\
M_{x\varphi} &= \frac{K_{x\varphi}}{a^2}w'^{\cdot}
\end{aligned}
\tag{7.1}
$$

Note: $(\ )' = a\dfrac{\partial(\ )}{\partial x}$; $(\ )^{\cdot} = \dfrac{\partial(\ )}{\partial\varphi}$

- Case 2: sandwich type shells with various internal structures

$$
\begin{aligned}
N_\varphi &= \frac{D_\varphi}{a}(v^{\cdot}+w)-\frac{S_\varphi}{a^2}(w+w^{\cdot\cdot})+\frac{K_\varphi}{a^3}(w+w^{\cdot\cdot})+\frac{\nu D}{a}u'+\frac{\nu S}{a^2}w''\\
N_x &= \frac{D_x}{a}u'-\frac{S_x}{a^2}w''+\frac{\nu D}{a}(v^{\cdot}+w)+\frac{\nu S}{a^2}(v^{\cdot}-w^{\cdot\cdot})\\
N_{\varphi x} &= \frac{(1-\nu)}{2a}D(u^{\cdot}+v')-\frac{(1-\nu)}{2a^2}S(u^{\cdot}-v'+2w'^{\cdot})+\frac{1-\nu}{2a^3}K(u^{\cdot}+w'^{\cdot})\\
N_{x\varphi} &= \frac{(1-\nu)}{2a}D(u^{\cdot}+v')+\frac{(1-\nu)S}{a^2}(v'-w'^{\cdot})+\frac{1-\nu}{2a^3}K(v'-w'^{\cdot})
\end{aligned}
$$

$$\begin{aligned}
M_{\varphi} &= -\frac{S_{\varphi}}{a}(v^{\cdot}+w)+\frac{K_{\varphi}}{a^2}(w+w^{\cdot\cdot})-\frac{\nu S}{a}u'+\frac{\nu K}{a^2}w'' \\
M_{x} &= -\frac{S_{x}}{a}u'+\frac{K_{x}}{a^2}w''-\frac{\nu S}{a}(v^{\cdot}+w)-\frac{\nu K}{a^2}(v^{\cdot}-w^{\cdot\cdot}) \\
M_{\varphi x} &= -\frac{(1-\nu)S}{2a}(u^{\cdot}+v')+\frac{(1-\nu)}{2a^2}K(u^{\cdot}-v'+2w'^{\cdot}) \\
M_{x\varphi} &= -\frac{(1-\nu)}{2a}S(u^{\cdot}+v')-\frac{(1-\nu)}{a^2}K(v'-w'^{\cdot})
\end{aligned} \tag{7.2}$$

- Case 3: solid shells made of high-strength orthotropic material

$$\begin{aligned}
N_{\varphi} &= \frac{D_{\varphi}}{a}(v^{\cdot}+w)+\frac{D_{\nu}}{a}u'+\frac{K_{\varphi}}{a^3}(w+w^{\cdot\cdot}) \\
N_{x} &= \frac{D_{x}}{a}u'+\frac{D_{\nu}}{a}(v^{\cdot}+w)-\frac{K_{x}}{a^3}w'' \\
N_{\varphi x} &= \frac{D_{\varphi x}}{a}(u^{\cdot}+v')+\frac{K_{\varphi x}}{a^3}(u^{\cdot}+w'^{\cdot}) \\
N_{x\varphi} &= \frac{D_{x\varphi}}{a}(u^{\cdot}+v')+\frac{K_{x\varphi}}{a^3}(v'-w'^{\cdot}) \\
M_{\varphi} &= \frac{K_{\varphi}}{a^2}(w+w^{\cdot\cdot})+\frac{K\nu}{a^2}w'' \\
M_{x} &= \frac{K_{x}}{a^2}(w''-u')+\frac{K\nu}{a^2}(w^{\cdot\cdot}-v^{\cdot}) \\
M_{\varphi x} &= \frac{K_{x\varphi}}{a^2}(2w'^{\cdot}+u^{\cdot}-v') \\
M_{x\varphi} &= \frac{2K_{x\varphi}}{a^2}(w'^{\cdot}-v')
\end{aligned} \tag{7.3}$$

- Case 4: isotropic shells

$$\begin{aligned}
N_{\varphi} &= \frac{D}{a}(v^{\cdot}+w+\nu u')+\frac{K}{a^3}(w+w^{\cdot\cdot}) \\
N_{x} &= \frac{D}{a}(u'+\nu v^{\cdot}+\nu w)-\frac{K}{a^3}w'' \\
N_{\varphi x} &= \frac{D}{a}\,\frac{(1-\nu)}{2}(u^{\cdot}+v')+\frac{K}{a^3}\,\frac{(1-\nu)}{2}(u^{\cdot}+w'^{\cdot}) \\
N_{x\varphi} &= \frac{D}{a}\,\frac{(1-\nu)}{2}(u^{\cdot}+v')+\frac{K}{a^3}\,\frac{(1-\nu)}{2}(v'-w'^{\cdot})
\end{aligned} \tag{7.4}$$

$$M_\varphi = \frac{K}{a^2}(w + w^{\cdot\cdot} + \nu w'')$$

$$M_x = \frac{K}{a^2}(w'' + \nu w^{\cdot\cdot} - u' - \nu v^{\cdot})$$

$$M_{\varphi x} = \frac{K}{a^2}(1-\nu)\,(w'^{\cdot} + \frac{1}{2}u' - \frac{1}{2}v')$$

$$M_{x\varphi} = \frac{K}{a^2}(1-\nu)\,(w'^{\cdot} - v')$$

A general elastic law can be written that encompasses all these cases:

$$N_\varphi = \frac{D_\varphi}{a}(\nu^{\cdot} + w) + \frac{D_\nu}{a}u' - \frac{S_\varphi}{a^2}(\alpha w + w^{\cdot\cdot}) + \bar{\rho}\frac{K_\varphi}{a^3}(w + w^{\cdot\cdot}) - \frac{\nu S}{a^2}w''$$

$$N_x = \frac{D_x}{a}u' + \frac{D_\nu}{a}(v^{\cdot} + w) - \frac{S_x}{a^2}w'' - \bar{\xi}\frac{K_x}{a^3}w'' + \frac{\nu S}{a^2}(v^{\cdot} - w^{\cdot\cdot})$$

$$N_{\varphi x} = \frac{D_{\varphi x}}{a}(u^{\cdot} + v') + \bar{\eta}\frac{K_{\varphi x}}{a^3}(\beta u^{\cdot} + w'^{\cdot}) - \frac{(1-\nu)}{2a^2}S(u^{\cdot} - v' + 2w'^{\cdot})$$

$$N_{x\varphi} = \frac{D_{x\varphi}}{a}(u^{\cdot} + v') + \bar{\epsilon}\frac{K_{x\varphi}}{a^3}(v' - w'^{\cdot}) + \frac{(1-\nu)}{a^2}S(v' - w'^{\cdot}) \tag{7.5}$$

$$M_\varphi = \frac{K_\varphi}{a^2}(\gamma w + w^{\cdot\cdot}) + \frac{K\nu}{a^2}w'' - \frac{S_\varphi}{a}(v^{\cdot} + w) - \frac{\nu S}{a}u'$$

$$M_x = \frac{K_x}{a^2}(w'' + \delta u') + \frac{K\nu}{a^2}(w^{\cdot\cdot} - \xi v^{\cdot}) - \frac{S_x}{a}\rho u' - \frac{\nu S}{a}(\nu^{\cdot} + w)$$

$$M_{\varphi x} = \frac{K_{\varphi x}}{a^2}(2w'^{\cdot} + \bar{\alpha}u^{\cdot} - \bar{\beta}v') - \frac{(1-\nu)}{2a}S(u^{\cdot} + v')$$

$$M_{x\varphi} = \bar{\delta}\frac{K_{x\varphi}}{a^2}(w'^{\cdot} + \bar{\gamma}v') - \frac{(1-\nu)}{2a}S(u^{\cdot} + v')$$

Each case is then found by substituting the appropriate values for the constants:

- Case 1

$$\alpha = 0,\ \bar{\rho} = 0,\ \bar{\xi} = 0,\ \bar{\eta} = 2,\ \beta = 0,\ \bar{\epsilon} = 0,\ \gamma = 0,\ \delta = 0,$$

$$\xi = 0,\ \rho = 1,\ \bar{\alpha} = 0,\ \bar{\beta} = 0,\ \bar{\gamma} = 0,\ \bar{\delta} = 1$$

$$D_\varphi = \frac{Et}{1-\nu^2} + \frac{EA_\varphi}{b_1} \qquad D_x = \frac{Et}{1-\nu^2} + \frac{EA_x}{b_2}$$

$$D_\nu = \frac{Et\nu}{1-\nu^2} \qquad D_{x\varphi} = \frac{Et}{2(1+\nu)}$$

$$S_\varphi = \frac{EA_\varphi C_\varphi}{b_1} \qquad S_x = \frac{EA_x C_x}{b_2}$$

$$K_\varphi = \frac{Et^3}{12(1-\nu^2)} + \frac{E(I_\varphi + A_\varphi C_\varphi^2)}{b_1} \qquad S = 0$$

$$K_x = \frac{Et^3}{12(1+\nu^2)} + \frac{E(I_x + A_x C_x^2)}{b_2}$$

$$K_v = \frac{Et^3 v}{12(1-v^2)}$$

$$K_{\varphi x} = \left[\frac{Et^3}{12(1+\nu)} + \frac{G J_\varphi}{b_1}\right]\frac{1}{2} \qquad K_{x\varphi} = \frac{Et^3}{12(1+\nu)} + \frac{G J_x}{b_2} \tag{7.6}$$

where
A_x, A_φ are cross-sectional areas of stiffener
C_x, C_φ are distances from neutral plane of stiffener to neutral plane of shell
I_x, I_φ are moments of inertia of stiffeners about their own neutral planes
$G J_x$, $G J_\varphi$ are torsional stiffnesses of stiffeners in the two directions

- Case 2

$$\alpha = 1,\ \bar{\rho} = 1,\ \bar{\xi} = 0,\ \beta = 1,\ \bar{\epsilon} = 1,\ \gamma = 1,\ \delta = 0,\ \bar{\eta} = 1$$

$$\xi = 1,\ \rho = 1,\ \bar{\alpha} = 1,\ \bar{\beta} = 1,\ \bar{\gamma} = -1,\ \bar{\delta} = 2$$

$$D_\varphi = \frac{E}{1-\nu^2}(t_1 + t_2) + \frac{EA_\varphi'}{b_1} \qquad D_\nu = \frac{\nu E(t_1 + t_2)}{1-\nu^2}$$

$$D_x = \frac{E}{1-\nu^2}(t_1 + t_2) + \frac{EA_x'}{b_2}$$

$$S_\varphi = \frac{E}{1-\nu^2}[C_\varphi' t_1 + C_\varphi'' t_2] + \frac{EA_\varphi'\bar{C}_\varphi}{b_1}$$

$$S_x = \frac{E}{1-\nu^2}[C_\varphi' t_1 + C_\varphi'' t_2] + \frac{EA_x'\bar{C}_\varphi}{b_2} \qquad \bar{C}_\varphi = \frac{C_\varphi' C_\varphi''}{2}$$

$$S = \frac{E}{1-\nu^2}[C_\varphi{}'t_1 + C_\varphi{}''t_2]$$

$$K = \frac{E}{3(1-\nu^2)}\left[\left(C_\varphi{}' + \frac{t_1}{2}\right)^3 - \left(C_\varphi{}' - \frac{t_1}{2}\right)^3 + \left(C_\varphi{}'' + \frac{t_2}{2}\right)^3 - \left(C_\varphi{}'' - \frac{t_2}{2}\right)^3\right]$$

$$K_\varphi = K + \frac{E\,I_\varphi{}'}{b_1}$$

$$K_x = K + \frac{E\,I_x{}'}{b_2}$$

$$D_{\varphi x} = \frac{(1-\nu)}{2}\frac{E(t_1+t_2)}{1-\nu^2} \qquad D_{x\varphi} = \frac{(1-\nu)}{2}\frac{E(t_1+t_2)}{1-\nu^2}$$

$$K_{x\varphi} = K_{\varphi x} = \frac{1-\nu}{2}\left[\left(C_\varphi{}' + \frac{t_1}{2}\right)^3 - \left(C_\varphi{}' \frac{t_1}{2}\right)^3 + \left(C_\varphi{}'' + \frac{t_2}{2}\right)^3 - \left(C_\varphi{}'' - \frac{t_2}{2}\right)^3\right]$$

$$K_\nu = \nu\left[\left(C_\varphi{}' + \frac{t_1}{2}\right)^3 - \left(C_\varphi{}' + \frac{t_1}{2}\right)^3 + \left(C_\varphi{}'' + \frac{t_2}{2}\right)^3 - \left(C_\varphi{}'' + \frac{t_2}{2}\right)^3\right] \tag{7.7}$$

where $I_x{}', I_\varphi{}'$ are moments of inertia of the internal stiffeners about their own neutral planes

- Case 3

$$\bar{\rho} = 1,\ \bar{\xi} = 1,\ \beta = 1,\ \bar{\eta} = 1,\ \gamma = 1,\ \bar{\epsilon} = 1,\ \rho = 0,\ \delta = -1$$

$$\xi = 1,\ \bar{\alpha} = 1,\ \bar{\beta} = 1,\ \bar{S} = 2,\ \bar{\gamma} = -1,\ \alpha = 0$$

$$D_x = E_1 t' \qquad D_\nu = E_\nu t'$$

$$D_\varphi = E_2 t' \qquad K_{x\varphi} = K_{\varphi x} = \frac{Gt'^3}{12}$$

$$D_{x\varphi} = D_{\varphi x} = Gt' \quad S_\varphi = 0$$

$$K_x = \frac{E_1 t'^3}{12} \qquad S = 0$$

$$S_x = 0$$

$$K_\varphi = \frac{E_2 t'^3}{12} \qquad K_\nu = \frac{E_\nu t'^3}{12} \tag{7.8}$$

Note: The orthotropic material is described by the stress–strain law:

$$\sigma_x = E_1 \in_x + E_\nu \in_\varphi, \quad \sigma_\varphi = E_\nu \in_x + E_2 \in_\varphi, \quad \tau_{x\varphi} = G\gamma_{x\varphi}$$

- Case 4

$$
\begin{aligned}
D_x &= \frac{E}{1-\nu^2} t'' & D_\nu &= \frac{\nu E t''}{1-\nu^2} \\
D_\varphi &= \frac{E}{1-\nu^2} t'' & K_{x\varphi} &= K_{\varphi x} = \frac{G t''^3}{12} \\
D_{x\varphi} &= D_{\varphi x} = G t'' & S_\varphi &= 0 \\
K_x &= \frac{E}{1-\nu^2} \frac{t''^3}{12} & S &= 0 \\
K_\varphi &= \frac{E}{1-\nu^2} \frac{t''^3}{12} & S_x &= 0 \\
K_\nu &= \frac{\nu E}{1-\nu^2} \frac{t''^3}{12}
\end{aligned}
\tag{7.9}
$$

7.3 General Equations for Shell Vibrating in Fluid and Containing Fluid

The general equations of motion for the shell can be written as[5]

$$
\begin{aligned}
&\frac{1}{D}[a\, N'_x + a\, N_{\varphi x}^{\bullet} + p_x a^2 - pa(u^{\bullet\bullet} - w') - Pu''] = \frac{\mu a^2}{D}\frac{\partial^2 u}{\partial t^2} + \frac{K a^2}{D}\frac{\partial u}{\partial t} \\
&\frac{1}{D}[a\, N^{\bullet}_\varphi + a\, N''_{x\varphi} - M^{\bullet}_\varphi - M'_{x\varphi} + p_\varphi a^2 - pa(v^{\bullet\bullet} + w^{\bullet}) - pv''] = \frac{\mu a^2}{D}\frac{\partial^2 v}{\partial t^2} + \frac{k a^2}{D}\frac{\partial v}{\partial t} \\
&\frac{1}{D}[M^{\bullet\bullet}_\varphi + M'^{\bullet}_{x\varphi} + M'^{\bullet}_{x\varphi} + M''_x + aN_\varphi + pa(u' - v^{\bullet} + w^{\bullet\bullet}) + pw''] \\
&\qquad = -\frac{\mu a^2}{D}\frac{\partial^2 w}{\partial t^2} - \frac{K a^2}{D}\frac{\partial w}{\partial t} + \frac{p_r a^2}{D} + \frac{a^2}{D}(p_{f_i} - p_{f_o})
\end{aligned}
\tag{7.10}
$$

Substituting the general elastic law, the equations in terms of the displacements u, v, and w are

$$
\begin{aligned}
&\frac{D_x}{D}u'' + \frac{D_\nu}{D}(v^{\bullet\prime} + w') - \frac{S_x}{aD}w''' - \bar{\xi}\frac{K_x}{a^2 D}w''' \\
&\quad + \frac{S}{Da}\left[(\nu v^{\bullet\prime} - \nu w^{\bullet\bullet\prime}) - \frac{(1-\nu)}{2}(u^{\bullet\bullet} - v^{\prime\bullet} + 2w^{\prime\bullet\bullet})\right] \\
&+ \frac{D_{\varphi x}}{D}(u^{\bullet\bullet} + v^{\prime\bullet}) + \bar{\eta}\frac{K_{\varphi x}}{Da^2}(\beta u^{\bullet\bullet} + w^{\prime\bullet\bullet}) + \frac{p_x a^2}{D} - \frac{pa}{D}(u^{\bullet\bullet} - w') - \frac{P}{D}u'' \\
&\qquad = \frac{\mu a^2}{D}\frac{\partial^2 u}{\partial t^2} + \frac{Ka^2}{D}\frac{\partial u}{\partial t} \\
&\frac{D_\varphi}{D}(v^{\bullet\bullet} + w^{\bullet}) + \frac{D_\nu}{D}u^{\prime\bullet} - \frac{S_\varphi}{Da}(\alpha w^{\bullet} + w^{\bullet\bullet\bullet} - v^{\bullet\bullet} - w^{\bullet}) + \frac{K_\varphi}{Da^2}(\bar{\rho} w^{\bullet} + \bar{\rho} w^{\bullet\bullet\bullet} - \gamma w^{\bullet} - w^{\bullet\bullet\bullet}) \\
&+ \frac{S}{Da}\left[-\nu w^{\prime\prime\bullet} + (1-\nu)(v'' - w^{\prime\prime\bullet}) + \nu u^{\prime\bullet} + \frac{(1-\nu)}{2}(u^{\prime\bullet} + v'')\right] + \frac{D_{x\varphi}}{D}(u^{\prime\bullet} + v'') \\
&+ \frac{K_{x\varphi}}{Da^2}(\bar{\in} v'' - \bar{\in} w^{\prime\prime\bullet} - \bar{\delta} w^{\prime\prime\bullet} - \bar{\delta}\bar{\gamma} v'') - \frac{K_\nu}{Da^2}w^{\prime\prime\bullet} + \frac{p_\varphi a^2}{D} - \frac{pa}{D}(v^{\bullet\bullet} + w^{\bullet}) - \frac{P}{D}v'' \\
&\qquad = \frac{\mu a^2}{D}\frac{\partial^2 v}{\partial t^2} + \frac{Ka^2}{D}\frac{\partial v}{\partial t} \\
&\frac{K_\varphi}{Da^2}(\gamma w^{\bullet\bullet} + w^{\bullet\bullet\bullet\bullet} + \bar{\rho} w + \bar{\rho} w^{\bullet\bullet}) + \frac{K\nu}{Da^2}(w^{\prime\prime\bullet\bullet} + w^{\bullet\bullet\prime\prime} - \xi v^{\bullet\prime\prime}) \\
&\quad + \frac{S_\varphi}{Da}(-v^{\bullet\bullet\bullet} - w^{\bullet\bullet} - \alpha w - w^{\bullet\bullet}) \\
&+ \frac{S}{Da}\left[-\nu u^{\prime\bullet\bullet} - \frac{(1-\nu)}{2}(u^{\prime\bullet\bullet} + v^{\prime\prime\bullet}) - \frac{(1-\nu)}{2}(u^{\bullet\bullet\prime} + v^{\prime\prime\bullet}) - \nu(v^{\bullet\prime\prime} + w'') - \nu w''\right] \\
&\quad + \frac{K_{x\varphi}}{Da^2}(\bar{\delta} w^{\prime\prime\bullet\bullet} + \bar{\delta}\bar{\gamma} v^{\prime\prime\bullet}) \\
&+ \frac{K_{\varphi x}}{Da^2}(2w^{\prime\prime\bullet\bullet} + \bar{\alpha} u^{\bullet\bullet\prime} - \bar{\beta} v^{\prime\prime\bullet}) + \frac{Kx}{Da^2}(w'''' + \delta u''') - \frac{S_x}{Da}\rho u''' + \frac{D\varphi}{D}(v^{\bullet} + w) + \frac{D\nu}{D}u' \\
&+ \frac{pa}{D}(u' - v^{\bullet} + w^{\bullet\bullet}) + \frac{Pw''}{D} = -\frac{\mu a^2}{D}\frac{\partial^2 w}{\partial t^2} - \frac{Ka^2}{D}\frac{\partial w}{\partial t} + \frac{p_r a^2}{D} + \frac{a^2}{D}(p_{f_i} - p_{f_a})
\end{aligned}
\tag{7.11}
$$

Using the infinite shell impedance as employed in Junger,[1] Bleich and Baron,[2] Greenspon,[3] and Leonard and Hedgepeth[6] and assuming freely supported ends in the elastic sense, the displacement functions will be

$$u = A\cos n\varphi \cos\frac{\lambda x}{a}e^{i\omega t}$$

$$v = B\sin n\varphi \sin\frac{\lambda x}{a}e^{i\omega t} \qquad \lambda = \frac{m\pi a}{\ell} \tag{7.12}$$

$$w = C\cos n\varphi \sin\frac{\lambda x}{a}e^{i\omega t}$$

and the equations of motion will reduce to

$$A\left[-k_1\lambda^2 + \frac{n^2(1-\nu)}{2}k_5 - k_6n^2 - \bar{\eta}\beta k_7 n^2 + q_1n^2 + q_2\lambda^2 + \Omega^2 - \frac{Ka^2}{D}i\omega\right]$$
$$+ B\left[h_2n\lambda + \nu k_5 n\lambda + k_5\frac{(1-\nu)}{2}n\lambda + k_6n\lambda\right]$$
$$+ C\left[k_2\lambda + k_3\lambda^3 + \bar{\xi}k_4\lambda^3 + \nu k_5n^2\lambda + k_5(1-\nu)n^2\lambda - \bar{\eta}k_7n^2\lambda + q_1\lambda\right] = -\frac{p_xa^2}{D} \tag{7.13}$$

$$A\left[k_2n\lambda + k_5\nu n\lambda + k_5\frac{(1-\nu)}{2}n\lambda + k_6n\lambda\right]$$
$$+ B\left[-k_8n^2 - k_9n^2 - k_5(1-\nu)\lambda^2 - k_5\frac{(1-\nu)}{2}\lambda^2 - k_6\lambda^2 - k_{11}\,\bar{\epsilon}\,\lambda^2\right.$$
$$\left. + k_{11}\bar{\delta}\bar{\gamma}\lambda^2 + q_1n^2 + q_2\lambda^2 + \Omega^2 - \frac{Ka^2}{D}i\omega\right]$$
$$+ C\left[-k_8n + k_9\alpha n - k_9n^3 - k_9n - k_{10}\bar{\delta}n + k_{10}\bar{\delta}n^3 + k_{10}\gamma n - k_{10}n^3 - k_5\nu n\lambda^2\right.$$
$$\left. - (1-\nu)n\lambda^2k_5 - k_{11}\,\bar{\epsilon}\,n\lambda^2 - k_{11}\bar{\delta}n\lambda^2 - k_{12}n\lambda^2 + q_1n\right] = -\frac{p_\varphi a^2}{D}$$

$$A[k_5\nu n^2\lambda + k_5(1-\nu)n^2\lambda - k_7\bar{\alpha}n^2\lambda - k_4\delta\lambda^3 + k_3\rho\lambda^3 + k_2\lambda + q_1\lambda]$$
$$+ B[-k_{12}\xi n\lambda^2 - k_9n^3 - k_5(1-\nu)n\lambda^2 - \nu n\lambda^2k_5 + k_{11}\bar{\delta}\bar{\gamma}n\lambda^2 - k_7\bar{\beta}n\lambda^2 - k_8n + q_1n]$$
$$+ C\left[\gamma k_{10}n^2 - k_{10}\,n^4 - k_{10}\bar{\delta} + k_{10}\bar{\delta}n^2 - 2k_{12}n^2\lambda^2 - 2k_9n^2 + k_9\alpha - 2k_5\nu\lambda^2\right.$$
$$- k_{11}\bar{\delta}n^2\lambda^2 - k_7 2n^2\lambda^2 - k_4\lambda^4 - k_8 + q_1n^2 + q_2\lambda^2 + \Omega^2 - \frac{Ka^2}{D}i\omega$$
$$\left. + \chi' - i\theta' + \gamma'\right] = -\frac{p_ra^2}{D}$$

p_x, p_φ, and p_r are forces per unit area applied to the shell in the longitudinal, tangential, and radial directions, respectively, where

$$k_1 = \frac{D_x}{D},\ k_2 = \frac{D_v}{D},\ k_3 = \frac{S_x}{aD},\ k_4 = \frac{K_x}{a^2 D},\ k_5 = \frac{S}{Da},\ k_6 = \frac{D_{x\varphi}}{D},\ k_7 = \frac{K_{\varphi x}}{Da^2}$$

$$q_1 = \frac{pa}{D},\ q_2 = \frac{P}{D},\ k_8 = \frac{D_\varphi}{D},\ k_9 = \frac{S_\varphi}{Da},\ k_{10} = \frac{K_\varphi}{Da^2},\ k_{11} = \frac{K_{x\varphi}}{Da^2},\ k_{12} = \frac{K\nu}{Da^2}$$

$$q^2 = \frac{\mu a^2}{D}\omega^2, \gamma^1 = r_2 \Omega^2 \gamma_{mn}, D = D_x \tag{7.14}$$

in which

$$r_2 = \frac{\rho_i a}{\mu}$$

where

ρ_i = density of internal fluid
μ = mass per unit area of structure

$$If\ \psi' C_r / C_i > 1 \quad \gamma_{mn} = \frac{J_n(k_r a)}{k_r a\left[-J_{n+1}(k_r a) + \frac{n}{k_r a} J_n(k_r a)\right]}$$

$$If\ \psi' C_r / C_i < 1 \quad \gamma_{mn} = \frac{I_n(k_r' a)}{k_r' a\left[I_{n+1}(k_r' a) + \frac{n}{k_r' a} In(k_r' a)\right]}$$

where

$$k_r a = \lambda\sqrt{(\psi' C_r / C_i)^2 - 1};\ k_r' a = \lambda\sqrt{1 - (\psi' C_r / C_i)^2};\ \psi' = \frac{\Omega}{\lambda}\frac{C_a}{C_r}$$

$$C_a = \sqrt{\frac{D}{\mu}},\quad C_r = \sqrt{\frac{G}{\rho}} \tag{7.15}$$

where

C_o = velocity of sound in surrounding medium
C_i = velocity of sound in medium inside shell

and

$$\chi' = \gamma_1 \frac{\lambda^2}{\Omega}\left(\frac{C_o}{C_a}\right)^3 \left(M - \frac{\Omega}{\lambda}\frac{C_a}{C_o}\right)^2 \chi_{mn}$$

in which

$$r_1 = \frac{a\rho_o}{\mu}$$

where
ρ_o = density of the medium surrounding the shell

$$\text{If } |M - \psi' C_r/C_o| < 1 \quad \chi_{mn}(k_m' a) = \frac{\lambda \psi' {}^{C_r}\!/\!_{C_o} K_n(k_m' r)}{k_m' a \left[K_{m-1}(k_m' a) + \dfrac{n}{k_n' a} K_n(k_m' a) \right]} \text{ at } r \tag{7.16}$$

$$\text{If } |M - \psi' C_r/C_o| > 1 \quad \chi_{mn}(k_m a) =$$

$$= \frac{-\lambda\psi' C_r/C_o \left\{ \left[J_n(k_m r) \left(-J_{n+1}(k_m a) + \dfrac{n}{k_m a} J_n(k_m a) \right) \right] + \left[Y_n(k_m r) \left(-Y_{n+1}(k_m a) + \dfrac{n}{k_m a} \right) Y_n(k_m a) \right] \right\}}{k_m a \left\{ \left[-J_{n+1}(k_m a) + \dfrac{n}{k_m a} J_n(k_m a) \right]^2 + \left[-Y_{n+1}(k_m a) + \dfrac{n}{k_m a} Y_n(k_m a) \right]^2 \right\}} \text{ at } r = a \tag{7.17}$$

$$k_m a = \lambda \sqrt{(M - \psi' C_r/C_o)^2 - 1} \qquad k_m' a = \lambda \sqrt{1 - (M - \psi' C_r/C_o)^2}$$

where
M = Mach number of the flow outside the cylinder

$$\theta' = r_1 \frac{\lambda^2}{\Omega} \left(\frac{C_o}{C_a} \right)^3 \left(M - \frac{\Omega}{\lambda} \frac{C_a}{C_o} \right)^2 \theta_{mn}$$

$$\text{If } |M - \psi' C_r/C_o| < 1 \qquad \theta_{mn} = 0$$

$$\text{If } |M - \psi' C_r/C_o| > 1,\ M > \psi C_r/C_o \ \ \theta_{mn} = \theta_m'{}_n =$$

$$\frac{\lambda\psi' C_r/C_o \left\{ J_n(k_m r) \left[-Y_{n+1}(k_m a) + \dfrac{n}{k_m a} Y_n(k_m a) \right] - Y_n(k_m r) \left[-J_{n+1}(k_m a) + \dfrac{n}{k_m a} Y_n(k_m a) \right] \right\}}{(k_m a) \left\{ \left[-J_{n+1}(k_m a) + \dfrac{n}{k_m a} J_n(k_m a) \right]^2 + \left[-Y_{n+1}(k_m a) + \dfrac{n}{k_m a} Y_n(k_m a) \right]^2 \right\}} \text{ at } r = a$$

$$\text{If } |M - \psi' C_r/C_o| > 1, M < \psi' C_r/C_o \quad \theta_{mn} = -\theta_{mn}'$$

Thus far, only the effects of vibration of the cylindrical surface have been considered. The motions of this surface are of primary consideration in the beam, lobar, and radial type modes. For the axially symmetric longitudinal modes, the longitudinal displacement of the ends plays a great part in the motion. For these latter types of modes, it will be assumed that the ends of the cylinder act like circular piston end caps vibrating in the acoustic medium. The largest effect will be experienced when the cylinder vibrates in modes in which the two ends are moving out of phase; the cylinder would then be moving like an accordion. The generalized forces associated with these end cap motions will involve the impedance of the piston. This impedance will be denoted by

$$Z_c = \theta_c + i\chi_c \tag{7.18}$$

where θ_c represents the resistive component and χ_c the reactive component.

If these terms are included in the general equations, the forced vibration equations can be written as

$$\begin{aligned}
&A\left[\left(a_{11} + a_{11}'q_1 + a_{11}''q_2 + \Omega^2 + \Omega^2\bar{k}\frac{4r_3}{\pi} + \bar{k}4r_1\frac{C_o}{C_a}r_4\Omega\chi_c\right)\right.\\
&\quad\left.+i\left(b_{11} + i\bar{k}4r_1\frac{C_o}{C_a}r_4\Omega\theta_c\right)\right] + B[a_{12}] + C[a_{13} + a_{13}'q_1] = P'\\
&A[a_{21}] + B[(a_{22} + a_{22}'q_1 + a_{22}''q_2 + \Omega^2) + i\,b_{22}] + C[a_{23} + a_{23}'q_1] = Q\\
&A[a_{31} + a_{31}'q_1] + B[a_{32} + a_{32}'q_1]\\
&\qquad\qquad + C[(a_{33} + a_{33}'q_1 + a_{33}''q_2 + \Omega^2 + \chi' + \gamma') + i\,b_{33}] = R
\end{aligned} \tag{7.19}$$

where

$$a_{11} = -k_1\lambda^2 + \frac{n^2(1-\nu)}{2}k_5 - k_6n^2 - \bar{\eta}\beta k_7n^2$$

$$a_{12} = k_2n\lambda + \nu k_5n\lambda + k_5\frac{(1-\nu)}{2}n\lambda + k_6n\lambda$$

$$a_{13} = k_2\lambda + k_3\lambda^3 + \bar{\xi}k_4\lambda^3 + \nu k_5n^2\lambda + k_5(1-\nu)n^2\lambda - \bar{\eta}k_7n^2\lambda$$

$$a_{21} = k_2n\lambda + k_5\nu n\lambda + k_5\frac{(1-\nu)}{2}n\lambda + k_6n\lambda$$

$$a_{22} = -k_8n^2 - k_9n^2 - k_5(1-\nu)\lambda^2 - k_5\frac{(1-\nu)}{2}\lambda^2 - k_6\lambda^2 - k_{11}\in\lambda^2 + k_{11}\bar{\delta}\gamma\lambda^2$$

$$a_{23} = -k_8n + k_9\alpha n - k_9n^3 - k_9n - k_{10}\bar{\delta}n + k_{10}\bar{\delta}n^3 + k_{10}\gamma n - k_{10}n^3$$
$$-k_5\nu n\lambda^2 - (1-\nu)n\lambda^2k_5 - k_{11}\in n\lambda^2 - k_{11}\bar{\delta}n\lambda^2 - k_{12}n\lambda^2$$

$$a_{31} = k_5\nu n^2\lambda + k_5(1-\nu)n^2\lambda - k_7\bar{\alpha}n^2\lambda - k_4\delta\lambda^3 + k_3\rho\lambda^3 + k_2\lambda$$

$$a_{32} = -k_{12}\xi n\lambda^2 - k_9n^3 - k_5(1-\nu)n\lambda^2 - \nu n\lambda^2k_5 + k_{11}\bar{\delta}\gamma n\lambda^2 - k_7\bar{\beta}n\lambda^2 - k_8n$$

$$a_{33} = \gamma k_{10}n^2 - k_{10}n^4 - k_{10}\bar{\rho} + k_{10}\bar{\rho}n^2 - 2k_{12}n^2\lambda^2 - 2k_9n^2 + k_9\alpha - 2k_5\nu\lambda^2$$
$$- k_{11}\bar{\delta}n^2\lambda^2 - 2k_7n^2\lambda^2 - k_4\lambda^4 - k_8 \tag{7.20}$$

$$b_{11} = -\Omega\bar{S}$$
$$b_{22} = b_{11}$$
$$b_{33} = -\theta' - \Omega\bar{S} \tag{7.21}$$
$$\bar{S} = \frac{K_a}{\mu C_a} \qquad a_{11}' = n^2,\ a_{11}'' = \lambda^2,\ a_{13}' = \lambda,\ a_{22}' = n^2,\ a_{22}'' = \lambda^2$$
$$a_{23}' = n,\ a_{31}' = \lambda,\ a_{32}' = n,\ a_{33}' = n^2,\ a_{33}'' = \lambda^2$$
$$C_a = \sqrt{D/\mu}$$

D is a reference stiffness, which can be chosen at random; however, once chosen, all calculations are referred to this number:

$$P = \frac{p_xa^2}{D},\quad Q = \frac{p_\varphi a^2}{D},\quad R = \frac{p_ra^2}{D} \qquad \bar{K} = 1 \text{ if } n = 0$$
$$\bar{K} = 0 \text{ if } n > 0$$

$$r_3 = \frac{\bar{M}}{\mu a\ell} \qquad r_4 = a/\ell \qquad \bar{M} = \textit{End cap mass}$$

7.4 Frequencies and Mode Shapes

The natural frequencies and mode shapes (ratio of constants) in vacuum can be obtained directly from the general equations. They are obtained by finding the eigenvalues and eigenvectors of the determinant:

$$\begin{vmatrix} (a_{11} + a_{11}'q_1 + a_{11}''q_2) + \Omega^2 & a_{12} & a_{13} + a_{13}'q_1 \\ a_{21} & (a_{22} + a_{22}'q_1 + a_{22}''q_2) + \Omega^2 & a_{23} + a_{23}'q_1 \\ a_{31} + a_{31}'q_1 & a_{32} + a_{32}'q_1 & (a_{33} + a_{33}'q_1 + a_{33}''q_2) + \Omega^2 \end{vmatrix} = 0 \tag{7.22}$$

The values of Ω that satisfy the preceding equation for a given set of physical parameters and modal configuration define the natural frequencies of the shell.

7.5 Effect of External or Internal Pressure on Natural Frequency

The determinant equation can be rewritten as

$$\begin{vmatrix} (a_{11}+n^2q_1+\lambda^2q_2)+\Omega^2 & a_{12} & a_{13}+\lambda q_1 \\ a_{21} & (a_{22}+n^2q_1+\lambda^2q_2)+\Omega^2 & a_{23}+nq_1 \\ a_{31}+\lambda q_1 & a_{32}+nq_1 & (a_{33}+n^2q_1+\lambda^2q_2)+\Omega^2 \end{vmatrix} = 0 \tag{7.23}$$

For most practical cases,

$$\begin{aligned} a_{13}, a_{31} &\gg \lambda q_1 \\ a_{23}, a_{32} &\gg nq_1 \end{aligned} \tag{7.24}$$

Let $\bar{\Omega}$ be defined as follows:

$$\bar{\Omega}^2 = \Omega + n^2q_1 + \lambda^2q_2$$

Then, the determinant reduces to

$$\begin{vmatrix} a_{11}+\bar{\Omega}^2 & a_{12} & a_{13} \\ a_{21} & a_{22}+\bar{\Omega}^2 & a_{23} \\ a_{31} & a_{32} & a_{33}+\bar{\Omega}^2 \end{vmatrix} = 0 \tag{7.25}$$

It is seen from Equation 7.23 that the $\bar{\Omega}$'s are the frequency numbers of the unpressureized shell (i.e., they are the values of Ω when $q_1 = q_2 = 0$). Thus, the circular frequencies ω_{mn} of the pressurized shell are given by

$$\omega_{mn} = \sqrt{\bar{\omega}_{mn}^{\;2} - \frac{(n^2pa + \lambda^2 P)}{\mu a^2}} \tag{7.26}$$

where ω_{mn} is the circular frequency of the pressurized shell and $\bar{\omega}_{mn}$ is the circular frequency of the unpressurized shell. Thus, for any static pressure, the frequency spectrum of a pressurized shell can be found from the spectrum of the unpressurized shell.

An equivalent relation to Equation 7.26 has been obtained by Fung, Sechler, and Kaplan[7] for a very thin, unstiffened shell. The previous relation (Equation 7.26) generalizes their results to thicker isotropic shells (i.e., shells for which the Flugge equations hold) and to those stiffened and sandwich shells that can be put into the category of equivalent anisotropic shells. According to Equation 7.26, an internal pressure will always give an increase in natural frequency, and an external pressure will always result in a decrease, no matter what mode is considered. This is in variance with the results of Baron and Bleich[2] for $n = 1$; however, their results approach the values predicted by this theory for $n > 1$.

The work of Arnold and Warburton[8] on fixed shells can be employed to generalize the results for other than freely supported boundary conditions (i.e., an equivalent λ can be employed in the case of a fixed or free shell).

For the case of a uniform external pressure on the shell,

$$P \approx pa/2 \tag{7.27}$$

Therefore, for this case,

$$\omega_{mn} = \sqrt{\bar{\omega}_{mn} - \frac{p}{\mu a}\left(n^2 + \frac{\lambda^2}{2}\right)} \tag{7.28}$$

For a uniform internal pressure, the second term under the radical becomes negative, thus giving an increase in frequency due to internal pressure.

7.6 Relation to Buckling

Upon examination of Equation 7.26 or 7.28, the question immediately arises concerning the case when the pressure is large enough to make the quantity under the square root negative. It can be shown that the pressure required to buckle the shell is also the eigenvalue of a determinant given by Equation 7.25. Call this the eigenvalue for buckling $(\bar{\Omega}_B)_{mn}$. The combination of q_1 and q_2 that will give buckling is given by the following expression:

$$(\bar{\Omega}_B)_{mn} = [n^2 q_1 + \lambda^2 q_2]_B \tag{7.29}$$

For a given m and n, $(\bar{\Omega}_B)_{mn}$ is numerically equal to $(\bar{\Omega})_{mn}$, the frequency number of the unpressurized shell. It must be emphasized that $(\bar{\Omega}_B)_{mn}$ is not the actual eigenvalue for buckling. It is only the eigenvalue that gives instability of the shell in the shape corresponding to the given value of m and n. This is pointed out by Flugge.[4]

Before, we noted that the vibration frequency number for the pressurized shell is given by

$$\Omega^2 = \bar{\Omega}^2 - (n^2 q_1 + \lambda^2 q_2) \tag{7.30}$$

Thus,

$$\Omega^2 = [n^2 q_1 + \lambda^2 q_2]_B - [n^2 q_1 + \lambda^2 q_2] \tag{7.31}$$

It is seen that any combination of q_1 and q_2 that results in $[n^2q_1 + \lambda^2q_2]$ becoming greater than $[n^2q_1 + \lambda^2q_2]_B$ implies that buckling would have already occurred.[9] Thus, we see that Ω is always positive since it is only possible for it to become zero when buckling is reached. The common points concerning the problems of buckling and vibration of shells were also discussed by Galletly[10] in an unpublished memorandum and for beams and plates by Lurie[11] and Johnson and Goldhammer.[12]

7.7 Forced Vibration

The equations for forced vibration were given in the previous section. P, Q, and R are Fourier components of the external forces (i.e., it is assumed that the external forces have been expanded into the same series as the displacements). We usually will need the response function for

$$P = Q = 0 \qquad R = \varepsilon^{i\omega t} \sin\frac{m\pi x}{\ell}\cos n\varphi \tag{7.32}$$

It should be noted that the theory includes the effect of speed (Mach number, M) on the impedances and the possibility of having internal fluid is also considered. The Mach number effects were added in accordance with the equations of Leonard and Hedgepeth.[6]

The frequency response function for lateral displacement and fluid pressure, assuming only lateral force R, can then be obtained as follows:

$$\begin{aligned} A\bar{a}_{11} + B\bar{a}_{12} + C\bar{a}_{13} &= 0 \\ A\bar{a}_{21} + B\bar{a}_{22} + C\bar{a}_{23} &= 0 \\ A\bar{a}_{31} + B\bar{a}_{32} + C(\bar{a}_{33} - iG') &= 1 \end{aligned} \tag{7.33}$$

Then,

$$C = \frac{\begin{vmatrix} \bar{a}_{11} & \bar{a}_{12} & 0 \\ \bar{a}_{21} & \bar{a}_{22} & 0 \\ \bar{a}_{31} & \bar{a}_{32} & 1 \end{vmatrix}}{\begin{vmatrix} \bar{a}_{11} & \bar{a}_{12} & \bar{a}_{13} \\ \bar{a}_{21} & \bar{a}_{22} & \bar{a}_{23} \\ \bar{a}_{31} & \bar{a}_{32} & \bar{a}_{33} - iG' \end{vmatrix}} \tag{7.34}$$

$$= \frac{(\bar{a}_{11}\bar{a}_{22} - \bar{a}_{12}\bar{a}_{21})}{\{[\bar{a}_{11}(\bar{a}_{22}\bar{a}_{33} - \bar{a}_{23}\bar{a}_{32}) - \bar{a}_{12}(\bar{a}_{21}\bar{a}_{33} - \bar{a}_{23}\bar{a}_{31}) + \bar{a}_{13}(\bar{a}_{21}\bar{a}_{32} - \bar{a}_{22}\bar{a}_{31})] - iG'[\bar{a}_{11}\bar{a}_{22} - \bar{a}_{12}\bar{a}_{21}]\}}$$

$$= C_r + iC_i$$

The frequency response functions for deflection and field pressure can then be written:

a. Deflection in mnth mode due to lateral pressure of $(p_r)_{mn} \cos n\varphi \sin \frac{m\pi\chi}{\ell}$:

$$w_{mn} = \frac{(p_r)_{mn} a^2}{D} \frac{\rho_o a}{\mu} \frac{\rho}{\rho_o} \frac{C_o^2}{C_a^2} \frac{a}{\rho C_o^2} (C_r + iC_i)_{mn} \sin \frac{m\pi x}{\ell} \cos n\varphi \tag{7.35}$$

b. Near field pressure in cylindrical coordinates in mnth mode due to lateral pressure of $(p_r)_{mn} \cos n\varphi \sin m\pi x/\ell$:

$$(p_{nf}(r,\varphi,x,w))_{mn} = (p_r)_{mn} \frac{a\rho_o}{\mu} \frac{C_o^2}{C_a^2} \frac{wa}{C_o}$$
$$\{[C_r\theta_{mn}(r) - C_i\chi_{mn}(r)] + i[C_i\theta_{mn}(r) + C_r\chi_{mn}(r)]\} \sin \frac{m\pi x}{\ell} \cos n \tag{7.36}$$

c. The far field pressure (in spherical coordinates) in mnth mode due to lateral pressure of $(p_r)_{mn} \cos n\varphi \sin m\pi x/\ell$:

$$(p_{ff}(R,Q,\psi,\omega))_{mn} = (p_r)_{mn} \frac{a\rho_o}{\mu} \frac{C_o^2}{C_a^2} \frac{\omega a}{C_o} \frac{a}{R} \varepsilon^{ikR} f(\varphi,\psi)$$

$$f(\varphi,\psi) = \frac{-i\omega(C_r + iC_i) \frac{2\rho_o C_o}{\sin\psi} \cos n\varphi \in^{in\pi/2} \frac{\bar{f}_n(k\cos\psi)}{a}}{J_n'(ka\sin\psi) + Y_n'(ka\sin\psi)} \tag{7.37}$$

where

$$m \text{ odd} \qquad \frac{\bar{f}_m(k)}{a} = \left(\frac{2}{\lambda}\cos\frac{k\ell}{2}\right) \Big/ \left[1-\left(\frac{ka}{\lambda}\right)^2\right]$$

$$m \text{ even} \qquad \frac{\bar{f}_m(k)}{a} = -i\left(\frac{2}{\lambda}\sin\frac{k\ell}{2}\right) \Big/ \left[1-\left(\frac{ka}{\lambda}\right)^2\right], (p_r)_{mn} = \frac{2p_o}{\pi a\ell}\sin\frac{m\pi c}{\ell}$$

For a force of $P_o e^{i\omega t}$ applied in the radial direction,

$$\begin{aligned} w(a,\varphi,x,\omega) &= \sum_{m=1}^{\infty}\sum_{n=1}^{\infty} w_{mn}(a,\varphi,x,\omega)e^{i\omega t} \\ p_{nf}(r,\varphi,x,\omega) &= \sum_{m=1}^{\infty}\sum_{n=1}^{\infty} (p_{nf}(r,\varphi,x,\omega))_{mn}e^{i\omega t} \\ p_{ff}(R,\varphi,\psi,\omega) &= \sum_{m=1}^{\infty}\sum_{n=1}^{\infty} (p_{ff}(R,\varphi,\psi,\omega))_{mn}e^{i\omega t} \end{aligned} \tag{7.38}$$

7.8 Description of the Computer Program

It is assumed that a lateral sinusoidal point load is applied at any point along the shell. The input to the program consists of the following parameters:

C5 = ratio of damping to critical damping (structural only)
C2 = $a\rho_o/\mu$, where a = shell radius, ρ_o = density of acoustic medium, and μ = mass per unit area of shell
C1 = C_o/C_A, where C_o = sound velocity in medium, $C_A = \sqrt{D_x/\mu}$, where $D_x = Eh_x$ and h_x = equivalent thickness for stretching in the x-direction
C4 = a/ℓ length of shell = ℓ
R1 = ρ_o/ρ density of shell material = ρ
C3 = $\omega a/C_o$ radian frequency of excitation = ω
$p1, p2, p3$ indices in the following series: $\sum_{n=0}^{Q}\sum_{m=p1}^{p2} A_{mn}\sin(m\pi x/\ell)\cos n\varphi$; P1 = minimum value of m, P2 = maximum value of m, P3 = increment in m, where m = number of axial half waves in the mode pattern

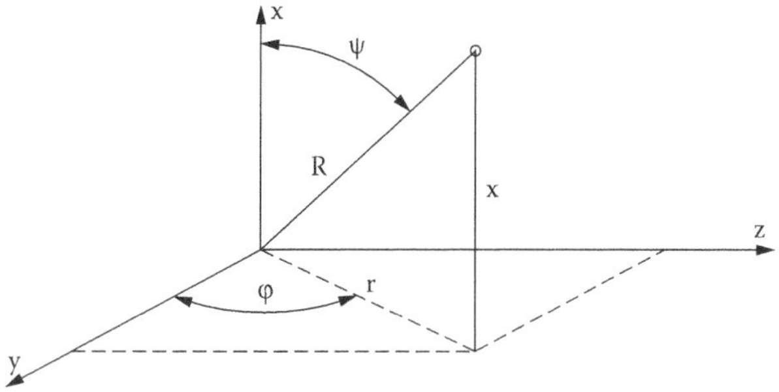

FIGURE 7.6
Coordinates.

M = an even integer greater than the largest of Ma
C8 = C/ℓ, where c is the x coordinate at which the lateral force is acting
W = peripheral coordinate φ (see Figure 7.6)
U1, U2, U3 = minimum, maximum, and increment of W
Q5, Q6, Q7 = minimum, maximum, and increment on C8
K9, L9 are scale factors that can be adjusted
S6 = θ angular coordinate in the spherical coordinate system
S7, S8, S9 = minimum, maximum, and increment on S6
D = Poisson's ratio
Q = largest n in the series $\sum_{n=0}^{Q} \sum_{m=P1}^{P2} A_{mn} \sin(m\pi x/\ell)\cos n\varphi$
Q9 = a constant value of 0.00001 (a program constant)
D1 = $\in/\ell$
E1 = δ (radians)
N = an even integer greater than 2 (C3)
N2 = an integer greater than C3, less than 2 (C3) (even integer)

The rest of the input is tied to the elastic structure, the theory of which is contained in Section 7.2 of this chapter.

Nomenclature in program	Nomenclature
B2	ρ
B3	α
B4	β
B5	γ
B6	δ
B7	ξ

M2	$\bar{\rho}$
M3	$\bar{\alpha}$
M4	$\bar{\beta}$
M5	$\bar{\gamma}$
M6	$\bar{\delta}$
M7	$\bar{\xi}$
M8	$\bar{\in}$
M9	$\bar{\eta}$
E4	k_1
E5	k_2
E6	k_3
E7	k_4
E8	k_5
E9	k_6
F4	k_7
F5	k_8
F6	k_9
F7	k_{10}
F8	k_{11}
F9	k_{12}

L8 = length of shell

A sample output of the program is shown in Table 7.1 and the numerical results of the program are shown in Table 7.2. An output print of "RES" as shown in the table illustrates that the forcing frequency is within the half-power points of the deflection amplitude at resonance. The other parameters are

S6 = polar angle in spherical coordinates

W = peripheral angle in spherical coordinates

W5 = imaginary part of $P_{ff}/(P_o/\pi a\ell)(a/R)\in^{ikR}$ in which P_{ff} is the far field pressure and R is the far field radial distance

W6 = real part of $P_{ff}/(P_o/\pi a\ell)(a/R)e^{ikR}$

W7 = absolute value of $P_{ff}/(P_o/\pi a\ell)(a/R)e^{ikR}$

C8 = C/1 where C = x coordinate at which lateral force is acting

I9 = far field sound pressure level in dynes/cm^2 per pound force at 1 yd

W9 = far field sound pressure level in dB/re .0002 dynes/cm^2 per pound force at 1 yd

TABLE 7.1

Computer Program for Point Loading of Cylindrical Shells

```
PROG4PR. BAS
10     DIM 10 DIM E (50, 50), F (50, 50)
30     DIM J (50), T (50), Y (50), L (50), H (50), K (50),
       N (50)
50     READ C5, C2, C1, C4, R1, C3, M, P1, P2, P3, Q5, Q6,
       Q7, D, K9, L9
60     READ L8
70     READ U1, U2, U3, S7, S8, S9, Q, Q9, D1, E1, N, N2
80     READ B2, B3, B4, B5, B6, B7
90     READ M2, M3, M4, M5, M6, M7, M8, M9
100    READ E4, E5, E6, E7, E8, E9
110    READ F4, F5, F6, F7, F8, F9
120    FOR I = 0 TO Q
130    IF I > 0 THEN 160
140    LET N(0) = .5
150    GOTO 170
160    LET N(I) = 1
170    NEXT I
180    FOR P=P1 TO P2 STEP P3
190    LET C6=P*3.14159*C4
210    IF C3>C6 THEN 3215
220    LET Z=SQR ((C6^2)-(C3^2))
230    GOSUB 3460
240    FOR I=0 TO N2
250    IF C3>C6 THEN 870
270    LET U4=0
280    LET A=Z*K(I+1)-I*K(I)
290    LET X4=(C3*K(I))/A
300    GOTO 980
870    LET B= (I/Z)*J(I)-J(I+1)
880    LET C= (I/Z)*Y(I)-Y(I+1)
890    IF ABS (J(I)/(10^K9))<ABS(Q9*Y(I)) THEN 950
900    LET D2=Z* ((B^2) + (10^(2*K9)) * (C^2))
910    LET U4=(C3*(10^K9)*(J(I)*C-Y(I)*B))/D2
920    LET 07=J (I)*B
930    LET X4=- (C3*(07+(10^(2*K9))*Y(I)*C))/D2
940    GOTO 980
950    LET U4=0
960    LET X4=- C3*Y(I)/(I*Y(I)-Z*Y(I+1))
980    GOSUB 5000
1450   NEXT I
1460   NEXT P
2600   FOR S6=S7 TO S8 STEP S9
```

(Continued)

TABLE 7.1 (Continued)

Computer Program for Point Loading of Cylindrical Shells

```
2610  LET G=SIN(S6)
2620  LET G1=C3*COS(S6)
2630  LET Z=C3*G
2635  LET M=N
2640  GOSUB 3220
2650  FOR C8=Q5 TO Q6 STEP Q7
2660  FOR W=U1 TO U2 STEP U3
2670  FOR P=P1 TO P2 STEP P3
2680  LET U7=SIN(P*1.5708)*COS(1.5708*(P-1))
2690  LET V7=COS(P*1.5708)*SIN(1.5708*(P+1))
2700  LET C6=P*3.1416*C4
2710  LET G2=(2/G)*(1/(1-(G1/C6)^2))
2720  LET G4=((2/C6)*SIN(G1/(2*C4)))*(1/6.2832)
2730  LET G3=((2/C6)*COS(G1/(2*C4)))*(1/6.2832)
2740  LET U9=U7*G2*G3
2750  LET V9=V7*G2*G4
2760  LET Z9=SIN (P*3.1416*C8)*(SIN(P*1.5708*E1))/
      (P*1.5708*E1)
2770  FOR I=0 TO N2
2780  LET E9=COS (1.5708*I)
2790  LET F9=SIN (1.5708*I)
2880  LET G7=(I/Z)*J(I)-J(I+1)
2810  LET G8=(I/Z)*Y(I)-Y(I+1)
2820  LET J1=E(P-P1+1, I) *E9+F (P-P1+1, I)*F9
2830  LET J2=F(P-P1+1, I) *E9-E(P-P1+1, I)*F9
2840  IF ABS (J(I)/(10^K9))<ABS(Q9*Y(I)) THEN 2910
2850  LET G9=(G7^2)+(10^(2*K9))*(G8^2)
2860  LET J3=U9*G7+V9*(10^K9)*G8
2870  LET J4=U9*(10^K9)*G8-V9*G7
2880  LET J5=(J1*J3-J2*J4)/G9
2890  LET J6=(J2*J3+J1*J4)/G9
2900  GOTO 2930
2910  LET J5=(J1*V9-J2*U9)/((10^K9)*G8)
2920  LET J6=(J2*V9+J1*U9)/((10^K9)*G8)
2930  LET P4=C2* (C1^2)*C3
2940  LET X9=N(I)*COS(I*W)* (SIN(.5*D1*I+.000001))/
      (.5*D1*I+.000001)
2950  LET Y9=X9*P4*Z9
2960  LET W5=W5+J5*Y9
2970  LET W6=W6+J6*Y9
3010  NEXT I
3020  NEXT P
3030  LET W7=SQR ((W5^2)+(W6^2))
```

TABLE 7.1 (Continued)

Computer Program for Point Loading of Cylindrical Shells

```
3050   LET I9=611*W7/L8
3060   LET W9=74+20* (LOG(I9))/(LOG(10))
3065   X=W7*COS (S6)
3067   Y=W7*SIN (S6)
3070   LPRINT "S6=";S6;"W=";W;"W5=";W5
3071   LPRINT "W6=";W6;"W7="W7;"C8=";C8
3090   LPRINT "I9=";I9;"W9=";W9
3092   LPRINT "X=";X;"Y=";Y
3100   LET W5=0
3110   I9=0
3120   W9=0
3130   LET W6=0
3140   LET W7=0
3180   NEXT W
3190   NEXT C8
3200   NEXT S6
3201   STOP
3202   DATA .05, 1.97, .706, .16, .127, 2, 20, 1, 21, 2, .5,
       .6, .7, .3, 0, 0
3204   DATA 756
3206   DATA 0, .1, .2, .1, 1.6, .1,6, .00001, .1, .1, 8, 6
3208   DATA 1, 0, 0, 0, 0, 0
3210   DATA 0, 0, 0, 0, 1, 0, 0, 2
3212   DATA 1, .219, 0, .00000466, 0, .28
3214   DATA .00000324, 1.36, .01725, .00052, .00000324,
       .00000139
3215   LET Z=SQR((C3^2)-(C6^2))
3216   GOSUB 3220
3217   GOTO 240
3220   LET T(M)=0
3230   LET T(M-1)=10^(-L9)
3240   FOR I=M TO 2 STEP -1
3250   LET T(I-2)=2*(I-1)*(1/Z)*T(I-1)-T(I)
3260   NEXT I
3270   LET S=0
3280   FOR I=1 TO M/2
3290   LET S=S+2*T (2*I)
3300   NEXT I
3310   LET S1=S+T(0)
3320   FOR 1=0 TO M
3330   LET J(I)=(T(I))/S1
3340   NEXT I
3350   LET S2=0
```

TABLE 7.1 (Continued)

Computer Program for Point Loading of Cylindrical Shells

```
3360  FOR I=1 TO M/2
3370  LET S2=S2+(2/I)*COS((I-1)*3.1416)*J(2*1)
3380  NEXT I
3390  LET Y(0)=((2/3.1416)*(J(0)*(LOG(.5*Z)+.577215665#)+
      S 2))/(10^K9)
3400  LET O9=Y(0)* (10^K9)
3410  LET Y(1)=((1/J(0))*(J(1)*O9-(2/(3.1416*Z))))/(10^K9)
3420  FOR I = 1 TO M
3430  LET Y(I+1)=2*I*(1/Z)*Y(I)-Y(I-1)
3440  NEXT I
3450  RETURN
3460  LET L(M)=0
3470  LET L(M-1)=10^(-L9)
3480  FOR I=M TO 2 STEP -1
3490  LET L(I-2)=2*(I-1)*(1/Z)*L(I-1)+L(I)
3500  NEXT I
3510  LET S3=0
3520  FOR I=1 TO M
3530  LET S3=S3+2*L(I)
3540  NEXT I
3550  LET S4=EXP(-Z)*(S3+L(0))
3560  FOR I=0 TO M
3570  LET H(I) = (L(I))/S4
3580  NEXT I
3590  IF Z<2 THEN 3610
3600  IF Z>2 THEN 3660
3610  LET U=.5*Z
3620  LET A=-LOG(.5*Z)*H(0)-.57721566#+.4227842#*U^2
3630  LET B=A+.23069756#*U^4+.0348859*U^6+2.62698E-03*U^8
3640  LET K(0)=(B+.0001075*U^10+.0000074*U^12)/(10^K9)
3650  GOTO 3710
3660  LET U=.5/Z
3670  LET V=(Z^.5)*EXP(Z)
3680  LET A=1.25331414#-7.832358E-02*U+2.189568E-02*U^2
3690  LET B=A-1.062446E-02*U^3+5.87872E-03*U^4
3700  LET K(0)=((B-.0025154*U^5+5.3208E-04*U^6)/V)/(10^K9)
3710  LET O8=K(0)*(10^K9)
3720  LET K(1) = ((1/H(0))*((1/Z)-08*H(1)))/(10^K9)
3730  FOR I=1 TO M
3740  LET K(I+1)=2*I*(1/Z)*K(I)+K(I-1)
3750  NEXT I
3760  RETURN
```

TABLE 7.1 (Continued)

Computer Program for Point Loading of Cylindrical Shells

```
5000  LET A0=(I^2)*(.5*(1-D)*E8-E9-M9*B4*F4)-
      E4*(C6^2)+((C3*C1)^2)
5005  LET K0=A0-(C3*C1)^2
5010  LET A1=I*C6*(E5+.5*(1+D)*E8+E9)
5015  LET A2=C6*(E5+(C6^2)*(E6+M7*E7)+(I^2)*(E8-M9*F4))
5020  LET A3=A1
5025  LET N4=(I^2)*(-F5-F6)+(C6^2)*(-1.5*(1-D)*E8-E9+(M6*M5-
      M8)*F8)
5030  LET A4=N4+((C3*C1)^2)
5035  LET N5=I*(-F5+(B3-1)*F6+(B5-M2)*F7)+(I^3)*(-F6+(M2-
      1)*F7)
5040  LET A5=N5+I*(C6^2)*(-E8-(M8+M6)*F8-F9)
5045  LET A6=C6*(E5+(C6^2)*(E6*B2-E7*B6)+(I^2)*C6*(E8-
      M3*F4))
5050  LET K4=N4
5055  LET N7= I*(-F5) + (I^3)*(-F6) +I*(C6^2)*(-B7*F9-
      E8+M6*M5*F8-M4*F4)
5060  LET A7=N7
5065  LET N8=(I^2)*((B5+M6)*F7-2*F6)-(I^4)*F7-(C6^4)*E7-
      2*(C6^2)*D*E8
5070  LET I8=N8+(I^2)*(C6^2)*(-2*F9-M6*F8-2*F4)-F7*M2+B3*F6
5075  A8=I8-F5+((C3*C1)^2)+C2*C3*(C1^2)*X4
5080  LET I3=A0*(A4*A8-A5*A7)-A1*(A3*A8-A5*A6)
5085  LET R3=I3+A2*(A3*A7-A4*A6)
5090  LET R4=A0*A4-A1*A3
5095  LET K8=I8-F5
5100  LET T1=K0*K4-A1*A3
5105  LET T2=K0*A7-A1*A6
5110  LET T3=A3*A7-K4*A6
5115  LET T4=(A5*T2-A2*T3)/T1
5120  LET T5=T4-K8
5125  LET T6=SQR(T5)
5130  LET R8=C2*C3*(C1^2)*U4+2*C5*C3*C1*T6
5135  LET R5=(R3^2)+(R4*R8)^2
5140  IF ABS(R3/R4)>R8 THEN 5160
5145  PRINT "RES";"I=";I;"P=";P
5150  LET F(P-P1+1, I) =(R4^2)*R8/R5
5155  GOTO 5165
5160  LET F(P-P1+1, I) =(R4^2)*R8/R5
5165  LET E(P-P1+1, I) =R4*R3/R5
5170  RETURN
7000  END
```

TABLE 7.2

Numerical Results from the Program Shown in Table 7.1

```
S6= .1 W= 0 W5= - .109851
W6= .8381396 W7= .8453077 C8 = .5
I9= .6831786 W9 = 70.69068
X= .8410847 Y= 8.438996E-02
S6= .2 W= 0 W5= .1127921
W6= .8666266 W7= .8739358 C8= .5
I9= .7063159 W9= 70.97998
X= .8565153 Y= .1736243
S6= .3 W= 0 W5 = .3048053
W6= .8036039 W7 = .8594681 C8= .5
I9= .6946231 W9= 70.83498
X= .8210813 Y= .2539902
S6= .4 W= 0 W5= .410703
W6= .6852151 W7= .7988721 C8= .5
I9= .6456493 W9= 70.19994
X= .73581 Y= .3110955
S6= .5 W= 0 W5 = .4170923
W6= .572263 W7 = .708132 C8= .5
I9= .572313 W9= 69.15267
X= .6214443 Y= .3394965
S6= .6 W= 0 W5= .3393352
W6= .4924101 W7= .5980101 C8= .5
I9= .4833124 W9= 67.68456
X= .493559 Y= .3376619
S6= .7000001 W= 0 W5= .1886768
W6= .4408207 W7= .4795016 C8= .5
I9= .3875337 W9= 65.76619
X= .366743 Y= .3089034
S6= .8000001 W= 0 W5= -2.911979E-02
W6= .3947786 W7= .3958511 C8= .5
I9= .3199273 W9 = 64.10102
X= .2757921 Y= .2839662
S6= .9000001 W= 0 W5= -.3069376
W6= .3226706 W7= .4453393 C8= .5
I9= .3599236 W9= 65.12421
X= .2768273 Y= .3488463
S6= 1 W= 0 W5 = -.632284
W6= .1941758 W7= .6614281 C8= .5
I9= .534567 W9= 68.56005
X= .3573711 Y= .5565727
S6= 1.1 W= 0 W5= -.9839256
W6= -5.643916E-03 W7= .9839418 C8= .5
I9= .7952228 W9= 72.00978
```

TABLE 7.2 (Continued)

Numerical Results from the Program Shown in Table 7.1

```
X= .4463121 Y= .8768962
S6= 1.2 W= 0 W5= -1.331598
W6= -.2664176 W7= 1.357988 C8= .5
I9= 1.097527 W9= 74.80831
X= .4920771 Y= 1.265698
S6= 1.3 W= 0 W5= -1.640181
W6= -.550689 W7= 1.73016 C8= .5
I9= 1.398317 W9= 76.91211
X= .4628155 Y= 1.66711
S6= 1.4 W= 0 W5= -1.876066
W6= -.8021336 W7= 2.040354 C8= .5
I9= 1.649016 W9= 78.3445
X= .3467926 Y= 2.010666
S6= 1.5 W= 0 W5= -2.013614
W6= -.9626798 W7= 2.231904 C8= .5
I9= 1.803827 W9= 79.1239
X= .1578781 Y= 2.22
```

7.9 Far Field Patterns for a Representative Case

Calculations were made for a cross-stiffened cylindrical shell having the proportions of a typical structure. The parameters used are shown in statements 3202–3214 of the program listing, except that the C3 ($\omega a/C_0 = ka$) was varied from 0.1 to 2.5. The resulting far field directivity patterns are shown in Figures 7.7 and 7.8. Note the transition from a dipole to a quadrapole type pattern as the *ka* goes from 0.1 to 1.0 and the persistency of the quadrapole behavior at *ka* = 2.0.

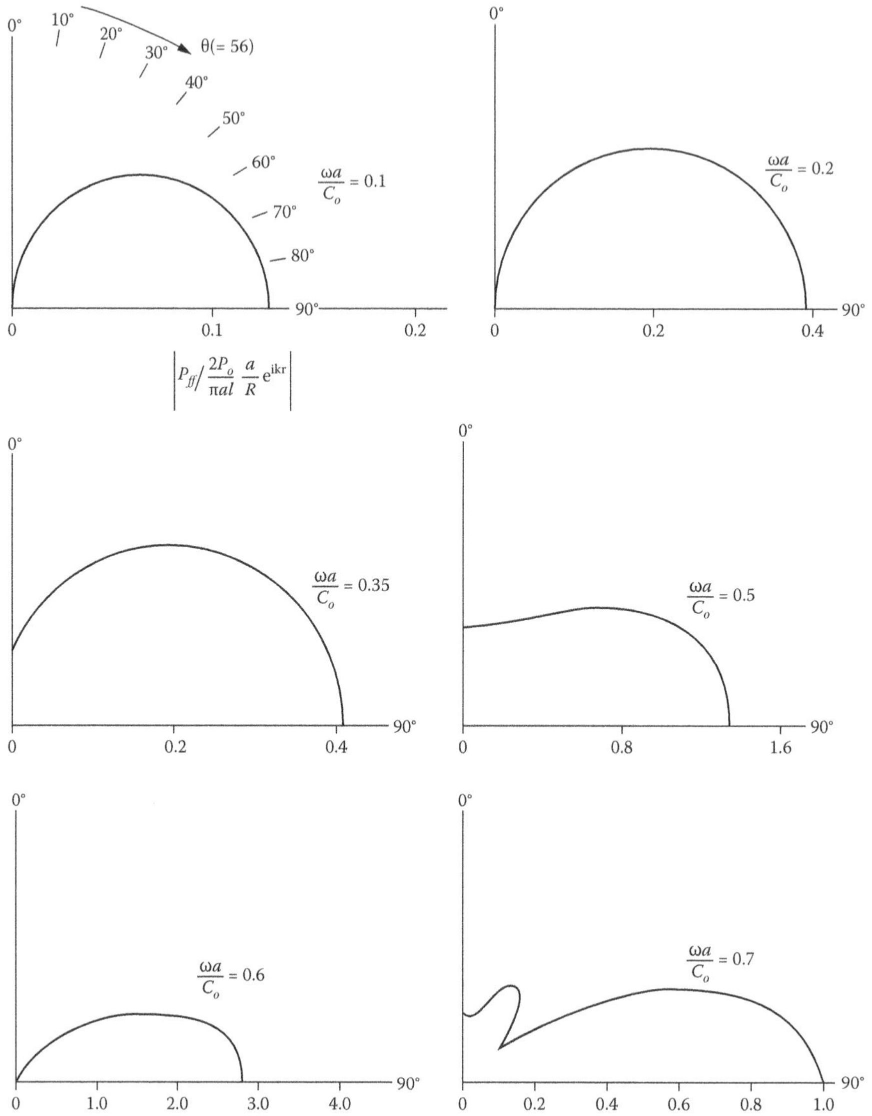

FIGURE 7.7
Far field directivity patterns for a stiffened cylindrical shell (axis of shell is X-axis) ($\varphi = 0$).

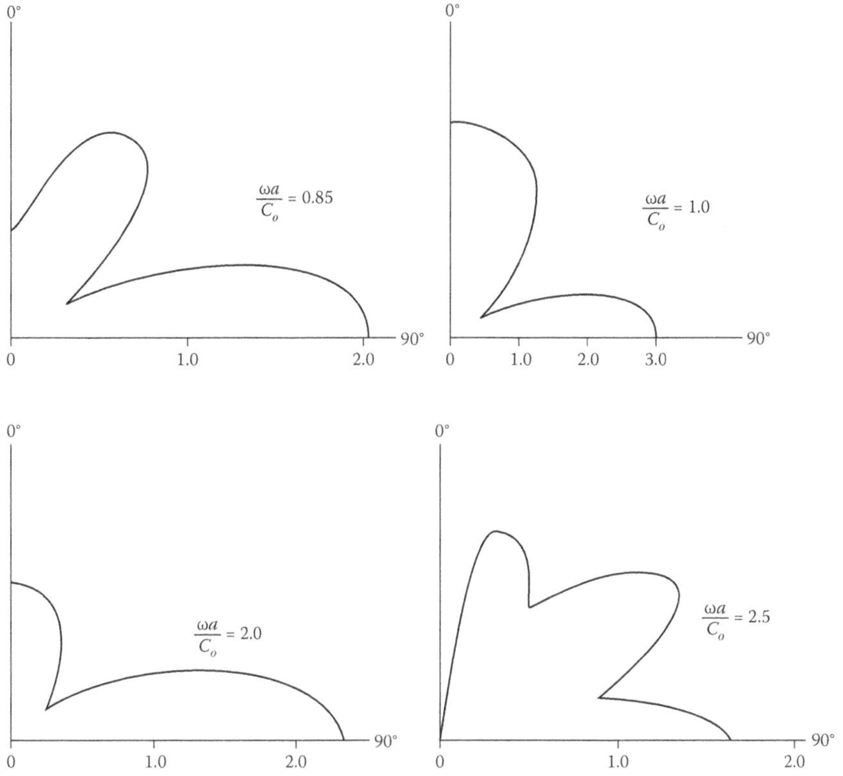

FIGURE 7.8
Far field directivity patterns for a stiffened cylindrical shell ($\varphi = 0$).

References

1. Junger, M. C. 1953. The physical interpretation of the expression for an outgoing wave in cylindrical coordinates. *Journal of the Acoustical Society of America* 25:40–47.
2. Bleich, H. H., and Baron, M. L. 1954. Free and forced vibrations of an infinitely long cylindrical shell in an infinite acoustic medium. *Journal of Applied Mechanics* 21:167–177.
3. Greenspon, J. E. 1961. Vibrations of thick and thin cylindrical shells surrounded by water. *Journal of the Acoustical Society of America* 33:1321–1328.
4. Flugge, W. 1962. *Stresses in shells,* 293. New York: Springer–Verlag.
5. Flugge, W. 1962. *Stresses in shells,* 422. New York: Springer–Verlag.
6. Leonard, R. W., and Hedgepeth, J. M. 1957. NACA, TR–1302.
7. Fung, Y. C., Sechler, E. E., and Kaplan, A. 1960. On the vibration of cylindrical shells under internal pressure. *Journal of Aerospace Science* 24:650–660.

8. Arnold, R. N., and Warburton, G. B. 1953. The flexural vibration of thin cylinders. *Proceedings of Institute of Mechanical Engineers* 167:62–64.
9. Flugge, W. 1962. *Stresses in shells,* 293, 424. New York: Springer–Verlag.
10. Galletly, G. D. 1953. Note on the relation between the vibration and general instability problems for stiffened cylindrical shells. Unpublished memorandum, David Taylor Model Basin, Washington, DC.
11. Lurie, H. 1952. Lateral vibrations as related to structural stability. *Journal of Applied Mechanics* 19:195–204.
12. Johnson, E. E., and Goldhammer, B. F. 1952. A determination of the critical load of a column on stiffened panel in compression by the vibration method. David Taylor Model Basin report 800.

8

Analysis of Three-Dimensional Media with Variable Properties

8.1 Introduction

Thus far in this book, we have analyzed beams, plates, and shells. This chapter will treat a three-dimensional medium that can have different properties throughout the medium. Part of the medium can be solid and part can be fluid, and the properties can be elastic and viscoelastic.

Any three-dimensional structure consisting of elastic, viscoelastic, and fluid portions can be analyzed with this model. The propagation of waves in the medium is analyzed by approximating it first by the characteristics of a uniform medium and then correcting the solution until it satisfies the actual medium.

8.2 Physical Characteristics of the Mathematical Model

The mathematical model of the complex will be divided into blocks with different elastic characteristics, as shown in Figure 8.1. The reflection and transmission of the waves produced in the various media are automatically taken into account by the representation of the entire structure as one complex, nonhomogeneous medium.

8.3 Differential Equations of the Complex Medium

The coupled differential equations of the system are given by Equation 8.1. These equations were manipulated so that all parts in the equations of motion are given in terms of displacements:

$$G_i = \frac{E_i}{2(1 + \mu_i)}, \quad \lambda_i = \frac{\mu_i E_i}{(1 + \mu_i)(1 - 2\mu_i)}$$

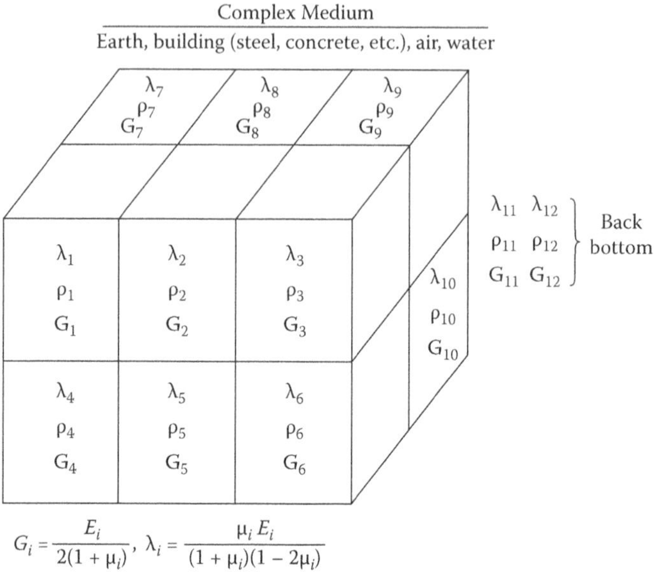

FIGURE 8.1

The mathematical model. λ_i, G_i = Lame's constants for the ith block; E_i = elastic modulus for the ith block; ρ_i = mass density of the ith block; μ_i = Poisson's ratio for the ith block.

$$\frac{\partial \lambda}{\partial x}\left(\frac{\partial u}{\partial x} + \frac{\partial v}{\partial y} + \frac{\partial w}{\partial z}\right) + \lambda\left(\frac{\partial^2 u}{\partial x^2} + \frac{\partial^2 v}{\partial x \partial y} + \frac{\partial^2}{\partial x \partial z}\right) + 2\frac{\partial G}{\partial x}\frac{\partial u}{\partial x} + 2G\frac{\partial^2 u}{\partial x^2}$$

$$+ \frac{\partial G}{\partial y}\left(\frac{\partial u}{\partial y} + \frac{\partial y}{\partial x}\right) + G\left(\frac{\partial^2 u}{\partial y^2} + \frac{\partial^2 v}{\partial y \partial x}\right)$$

$$+ \frac{\partial G}{\partial z}\left(\frac{\partial u}{\partial z} + \frac{\partial w}{\partial x}\right) + G\left(\frac{\partial^2 u}{\partial z^2} + \frac{\partial^2 w}{\partial z \partial x}\right)$$

$$+ X(x,y,z,t) - \frac{\partial}{\partial x}\left(\frac{\alpha ET}{1 - 2\bar{v}}\right) - \rho(x,y,z)\frac{\partial^2 u}{\partial t^2} = 0$$

$$\frac{\partial \lambda}{\partial y}\left(\frac{\partial u}{\partial x} + \frac{\partial v}{\partial y} + \frac{\partial w}{\partial z}\right) + \lambda\left(\frac{\partial^2 u}{\partial y \partial x} + \frac{\partial^2 v}{\partial y^2} + \frac{\partial^2 w}{\partial y \partial z}\right) + 2\frac{\partial G}{\partial y}\frac{\partial v}{\partial y} + 2G\frac{\partial^2 v}{\partial y^2}$$

$$+ \frac{\partial G}{\partial x}\left(\frac{\partial u}{\partial y} + \frac{\partial v}{\partial x}\right) + G\left(\frac{\partial^2 u}{\partial x \partial y} + \frac{\partial^2 v}{\partial x^2}\right)$$

$$+ \frac{\partial G}{\partial z}\left(\frac{\partial v}{\partial z} + \frac{\partial w}{\partial y}\right) + G\left(\frac{\partial^2 v}{\partial z^2} + \frac{\partial^2 w}{\partial z \partial y}\right)$$

$$+ Y(x,y,z,t) - \frac{\partial}{\partial y}\left(\frac{\alpha ET}{1 - 2\bar{v}}\right) - \rho(x,y,z)\frac{\partial^2 v}{\partial t^2} = 0$$

$$\frac{\partial\lambda}{\partial z}\left(\frac{\partial u}{\partial x}+\frac{\partial v}{\partial y}+\frac{\partial w}{\partial z}\right)+\lambda\left(\frac{\partial^2 u}{\partial z\partial x}+\frac{\partial^2 v}{\partial z\partial y}+\frac{\partial^2 w}{\partial z^2}\right)+2\frac{\partial G}{\partial z}\frac{\partial w}{\partial z}+2G\frac{\partial^2 w}{\partial z^2}$$
$$+\frac{\partial G}{\partial x}\left(\frac{\partial u}{\partial z}+\frac{\partial w}{\partial x}\right)+G\left(\frac{\partial^2 u}{\partial x\partial z}+\frac{\partial^2 w}{\partial x^2}\right)$$
$$+\frac{\partial G}{\partial y}\left(\frac{\partial v}{\partial z}+\frac{\partial w}{\partial y}\right)+G\left(\frac{\partial^2 v}{\partial y\partial z}+\frac{\partial^2 w}{\partial y^2}\right)$$
$$+Z(x,y,z,t)-\frac{\partial}{\partial z}\left(\frac{\bar{\alpha}ET}{1-2\bar{v}}\right)-\rho(x,y,z)\frac{\partial^2 w}{\partial t^2}=0 \tag{8.1}$$

and
$\lambda = \lambda(x,y,z)$
$G = G(x,y,z)$
$T = T(x,y,z,t)$
$E = E(x,y,z)$
$\bar{v} = \bar{v}(x,y,z)$
$\bar{\alpha} = \bar{\alpha}(x,y,z)$
λ = Lame's dilatation constant
G = shear modulus
T = temperature
E = Young's modulus
$\bar{v}$ = Poisson's ratio
$\bar{\alpha}$ = coefficient of linear thermal expansion

8.4 Approximate Solution

The Eason, Fulton, and Sneddon solution[3] for an infinite homogeneous isotropic body subject to a point harmonic load is given next.

Consider the distribution of stress produced by a point force of maximum magnitude F, pointing in the z-direction and varying harmonically with the time. If we take cylindrical coordinates r, z with origin at the point of application of the force, then we have

$$Z=\frac{F}{\rho}e^{iqr}\frac{\delta(z)\delta(r)}{2\pi r}\quad (q=p/c_1), \tag{8.2}$$

so that

$$\bar{Z}=\frac{F}{2\pi\rho}\delta(\omega+q) \tag{8.3}$$

If we substitute from Equation 8.3 into the equations given in Eason et al.[3] and perform the integration with respect to ω, which is immediate, we find that the components of the displacement vector may be written in the form

$$u = \frac{Fe^{iqr}}{4\pi^2\mu\beta^2q^2}\frac{\partial^2}{\partial r\partial z}\{I(\beta q) - I(q)\}, \tag{8.4}$$

$$w = \frac{Fe^{iqr}}{4\pi^2\mu\beta^2q^2}\left\{\beta^2q^2I(\beta q) + \frac{\partial^2}{\partial z^2}[I(\beta q) - I(q)]\right\}, \tag{8.5}$$

where we have adopted the notation

$$I(q) = \int_0^\infty \xi J_0(\xi r)d\xi \int_{-\infty}^{\infty}\frac{e^{-i\zeta z}d\zeta}{\xi^2 + \zeta^2 - q^2} \tag{8.6}$$

The ζ-integration in $I(q)$ may be effected by means of the calculus of residues and the resulting ξ-integration by means of a well-known result in the theory of Bessel functions. We find finally that, when the value

$$I(q) = \frac{\pi e^{-iq(r^2+z^2)}}{\sqrt{(r^2+z^2)}} \tag{8.7}$$

for $I(q)$ is inserted into Equations 8.4 and 8.5, the components of the displacement vector for this problem take the form

$$u = \frac{rze^{iqr}F}{4\pi\mu\beta^2q^2}\left\{\frac{3}{R^5}(e^{-i\beta qR} - e^{-iqR}) + \frac{3iq}{R^4}(\bar{\beta}e^{-i\beta qR} - e^{-iqR}) - \frac{q^2}{R^3}(\beta^2e^{-i\beta qR} - e^{-iqR})\right\}, \tag{8.8}$$

$$w = \frac{e^{iqr}F}{4\pi\mu\beta^2q^2}\left\{\frac{\beta^2q^2}{R}e^{-i\beta qR} + \frac{(2z^2 - r^2)}{R^5}(e^{-i\beta qR} - e^{-iqR}) + \frac{(2z^2 - r^2)iq}{R^4}(\beta e^{-i\beta qR} - e^{-iqR})\right.$$
$$\left. - \frac{zq^2}{R^3}(\beta^2e^{-i\beta qR} - e^{iqR})\right\}, \tag{8.9}$$

where $R^2 = r^2 + z^2$.

$$\tau = C_1t, \beta^2 = \frac{\lambda + 2\mu}{\mu} = \frac{C_1^2}{C_2^2}, q = {}^{p}\!/_{C_1}$$
$$C_1 = \sqrt{(\lambda + 2v)/\rho} \qquad C_2 = \sqrt{\mu/\rho} \tag{8.10}$$

where p = frequency.

The general case for a load with arbitrary orientation in rectangular coordinates can be obtained from the Eason et al.[3] solution by employing the following:

Actual deflections in X-, Y-, and Z-directions are *U*, *V*, and *W* (*u* and *w* are the radial and longitudinal deflections in cylindrical coordinates given by Eason et al.[3]). In order to use the Eason et al.[3] solution for load components in the Z-direction, leave *z* as is in the solution and replace *r* with $\sqrt{x^2+y^2}$.

For load components in the Y-direction, replace *z* with *y* in the Eason et al.[3] equations and replace *r* with $\sqrt{y^2+z^2}$.*

The U component for the force component in the Z direction is u $x/\sqrt{x^2+y^2}$

The U component for the force component in the Y direction is u $x/\sqrt{x^2+z^2}$

The U component for the force component in the X direction is w

The V component for the force component in the Z direction is u $y/\sqrt{x^2+y^2}$

The V component for the force component in the Y direction is w

The V component for the force component in the X direction is u$y/\sqrt{y^2+z^2}$

The W component for the force component in the Z direction is w

The W component for the force component in the Y direction is u $z/\sqrt{x^2+z^2}$

The W component for the force component in the X direction is u $z/\sqrt{y^2+z^2}$

(8.11)

Once the response is computed for harmonic loading, the response for random loading is

$$\bar{\bar{S}}\begin{pmatrix} j_x,\omega \\ j_y,\omega \\ j_z,\omega \end{pmatrix} = \sum_{k=1}^{K} \left| \bar{\bar{G}}\begin{pmatrix} j_x,k_x,\omega \\ j_y,k_y,\omega \\ j_z,k_z,\omega \end{pmatrix} \right|^2 \bar{\bar{S}}_L\begin{pmatrix} k_x,\omega \\ k_y,\omega \\ k_z,\omega \end{pmatrix} \tag{8.12}$$

where

$\bar{\bar{S}}(j_x,\omega)$ = spectrum of the response at point *j* in the x-direction at frequency ω

$\bar{\bar{G}}(j_x,k_x,\omega)$ = response at point *j* in the x-direction due to unit harmonic load at *k* in the x-direction at frequency ω

$\bar{\bar{S}}_L(k_x,\omega)$ = spectrum of the load at point *k* in the x-direction, etc.

* (For cases where the load is located at point x_0, y_0, z_0, instead of the origin, as in the Eason et al. relations, just replace *x*, *y*, and *z* with $(x - x_0)$, $(y - y_0)$, and $(z - z_0)$.)

The $\overline{\overline{G}}$s are calculated from the extension of Eason and colleagues' equations given in the previous sections.

A more general solution for point loading of an infinite homogeneous isotropic space is given by Love[4] and Stokes.[5] The Stokes–Love solution gives the relations for a point load with any time dependency, whereas the previous relations were for harmonic loading and the resulting Green's functions, which could be used in random loading situations. The Love relations are as follows:

$$\beta = \sqrt{(\lambda + 2G)/\rho}, b = \sqrt{G/\rho}, r = \sqrt{(x - x_0)^2 + (y - y_0)^2 + (z - z_0)^2}$$

where x_0, y_0, z_0 = location of point load $X(t)$. For point load in the x-direction,

$$U = \frac{1}{4\pi\rho}\frac{\partial^2}{\partial x^2}\left(\frac{1}{r}\right)\left[\int_{r/a}^{r/b} t^1 \chi(t - t^1)dt^1\right] + \frac{1}{4\pi\rho r}\left(\frac{\partial r}{\partial x}\right)^2 \left\{\frac{1}{a^2}\chi\left(t - \frac{r}{a}\right) - \frac{1}{b^2}\chi\left(t - \frac{r}{b}\right)\right\}$$
$$+ \frac{1}{4\pi\rho b^2 r}\chi\left(t - \frac{r}{b}\right)$$

$$V = \frac{1}{4\pi\rho}\frac{\partial^2}{\partial x \partial y}\left(\frac{1}{r}\right)\left[\int_{r/a}^{r/b} t^1 \chi(t - t^1)dt^1\right] + \frac{1}{4\pi\rho r}\frac{\partial r}{\partial x}\frac{\partial r}{\partial y}\left\{\frac{1}{a^2}\chi\left(t - \frac{r}{a}\right) - \frac{1}{b^2}\chi\left(t - \frac{r}{b}\right)\right\}$$

$$W = \frac{1}{4\pi\rho}\frac{\partial^2}{\partial x \partial z}\left(\frac{1}{r}\right)\left[\int_{r/a}^{r/b} t^1 \chi(t - t^1)dt^1\right] + \frac{1}{4\pi\rho r}\frac{\partial r}{\partial x}\frac{\partial r}{\partial z}\left\{\frac{1}{a^2}\chi\left(t - \frac{r}{a}\right) - \frac{1}{b^2}\chi\left(t - \frac{r}{b}\right)\right\} \tag{8.13}$$

For a point load in the y-direction, make the following changes in the preceding equations:

U becomes *V* and *x* becomes *y* in the first equation.
V becomes *U*, *x* becomes *y*, and *y* becomes *x* in the second equation.
W stays as *W*, *x* becomes *y*, and *z* stays as *z* in the third equation.

For a point load in the z-direction, make the following changes:

U becomes *W* and *x* becomes *z* in the first equation.
V stays as *V*, *x* becomes *z*, and *y* stays as *y* in the second equation.
W becomes *U*, *x* becomes *z*, and *z* becomes *x* in the third equation.

8.5 Solution for Nonhomogeneous Systems

This section shows how the approximate solution can be used to obtain a new solution that is much closer to the exact solution of Equation 8.1.

Write the differential equations in Equation 8.1 in finite difference form, using central differences (see Allen[6]). The derivatives involved in the equations are as follows:

$$\frac{\partial \lambda}{\partial x} = \frac{\lambda(x+DX,y,z)-\lambda(x-DX,y,z)}{2\ DX}$$

$$\frac{\partial u}{\partial x} = \frac{u(x+DX,y,z,t)-u(x-DX,y,z,t)}{2\ DX}$$

$$\frac{\partial v}{\partial y} = \frac{v(x,y+DY,z,t)-v(x,y-DY,z,t)}{2\ DY}$$

$$\frac{\partial w}{\partial z} = \frac{w(x,y,z+DZ,t)-w(x,y,z-DZ,t)}{2\ DZ}$$

$$\frac{\partial^2 u}{\partial x^2} = \frac{u(x+DX,y,z,t)+u(x-DX,y,z,t)-2u(x,y,z,t)}{DX^2}$$

$$\frac{\partial^2 v}{\partial x \partial y} = \frac{\begin{array}{c} v(x+DX,y+DY,z,t)-v(x-DX,y+DY,z,t) \\ +v(x-DX,y-DY,z,t)-v(x+DX,y-DY,z,t) \end{array}}{4\ DXDY}$$

$$\frac{\partial^2 w}{\partial x \partial z} = \frac{\begin{array}{c} w(x+DX,y,z+DZ,t)-w(x-DX,y,z+DZ,t) \\ +w(x-DX,y,z-DZ,t)-w(x+DX,y,z-DZ,t) \end{array}}{4\ DXDZ}$$

$$\frac{\partial G}{\partial x} = \frac{G(x+DX,y,z,t)-G(x-DX,y,z,t)}{2\ DX}$$

$$\frac{\partial G}{\partial y} = \frac{G(x,y+DY,z,t)-G(x,y-DY,z,t)}{2\ DY} \qquad \text{etc.}$$

All of the derivatives in all three equations of motion of Equation 8.1[1] are of the same form as the preceding expressions. It is only necessary to change the function and apply the *DX, DY,* and *DZ* to the appropriate terms. Use the iteration procedure as outlined in Scarborough[7] and Carnahan, Luther, and Wilkes.[8] Use the approximate solution as given in Greenspon[1] and Eason et al.[3] to evaluate the displacements at points away from *x, y, z,* and *t*. The only terms that will contain *u, v,* and *w* evaluated at *x, y, z, t* (i.e., not evaluated at *x* + *DX*, etc.) are the ones involving the second derivatives with respect to *x, y,* and *z*. All the remaining terms will contain deflections on either side of the point *x, y, z* at time *t*. Thus, the first equation of motion will give *u*(*x, y, z, t*) explicitly in terms of *u, v,* and *w* values at points other than *x, y,* and *z*. Likewise, the second equation of motion will give *v*(*x, y, z, t*) and the third will give *w*(*x, y, z, t*).

The iteration will take place in two steps. The first step will be to adjust the values of *DX, DY,* and *DZ*. The second step will be to use the calculated values of *u, v,* and *w* at different values of *x, y, z, t* to do a second iteration. This procedure will continue until the values of *u, v,* and *w* do not change with further iteration.

Several references[9–12] discuss these types of problems.

References

1. Greenspon, J. E. 2000. *Coupled fluid-structural phenomena treated with nonhomogeneous modeling*. JG Engineering Research Associates.
2. Oleinik, O. A., Shamaev, A. S., and Yosifian, G. A. 1992. *Mathematical problems in elasticity and homogenization*. Amsterdam: North Holland Publishing.
3. Eason, G., Fulton, J., and Sneddon, I. N. 1956. The generation of waves in an infinite elastic solid by variable body forces. *Philosophical Transactions A* 248:575–607.
4. Love, A. E. H. 1944. *A treatise on the mathematical theory of elasticity,* 305. New York: Dover Publications.
5. Stokes, G. G. 1849. Mathematical and physical papers, vol. 2, 243 (published as "On the Dynamical Theory of Diffraction," *Cambridge Philosophical Society Transactions* 9:1).
6. Allen, D. N. de G. 1962. Chapter 13 in *Handbook of engineering mechanics,* ed. W. Flugge. New York: McGraw–Hill.
7. Scarborough, J. B. 1962. *Numerical mathematical analysis,* 5th ed., 390. Baltimore, MD: Johns Hopkins University Press.
8. Carnahan, B., Luther, H. A., and Wilkes, S. O. 1969. *Applied numerical methods,* 488. New York: John Wiley & Sons, Inc.
9. Yang, B., and Tan, C. A. 1992. Transfer functions of one-dimensional distributed parameter systems. *Journal of Applied Mechanics* 59:1009–1014.
10. Reshef, M., Kostoff, D., and Edwards, M. 1988. Three-dimensional elastic modeling by the Fourier method. *Geophysics* 53:1184.
11. Beaudet, P. R. 1970. Elastic wave propagation in heterogeneous media. *Bulletin of Seismological Society of America* 60:769–784.
12. Virieux, J. 1986. P-SV wave propagation in heterogeneous media. *Geophysics* 51:889.

Section 2

Random Phenomena

9

Linear Systems Equations[8]

9.1 Impulse Response

Consider a system with n inputs and m outputs as shown in Figure 9.1. Each of the inputs and outputs is a function of time. The central problem lies in trying to determine the outputs or some function of the outputs in terms of the inputs or some function of them. Let any input $x(t)$ be divided into a succession of impulses, as shown in Figure 9.2. Let $h(t - \tau)$ be the response at time t due to a unit impulse at time τ. A unit impulse is defined as one in which the area under the input versus time curve is unity. Thus, if the base is $\Delta\tau$, the height of the unit impulse is $1/\Delta\tau$. Thus, $h(t - \tau)$ is the response per unit area (or per unit impulse at $t - \tau$).

The area (or impulse) is $x(\tau)\Delta\tau$. The response at time t is the sum of the responses due to all the unit impulses for all time up to t—that is, from $-\infty$ to t. But it is physically impossible for a system to respond to anything but past inputs; therefore,

$$h(t-\tau) = 0 \text{ for } \tau > t \tag{9.1}$$

Thus, the upper limit of integration can be changed to $+\infty$. By a simple change of variable, $\theta = t - \tau$, it can be demonstrated that

$$y(t) = \int_{-\infty}^{+\infty} h(\theta)x(t-\theta)\, d\theta \tag{9.2}$$

Since there are n inputs and m outputs, there must be one of these equations for each input and output. Thus, for the ith input and jth output,

$$y_{ij}(t) = \int_{-\infty}^{+\infty} h_{ij}(\tau)x_i(t-\tau)\, d\tau \tag{9.3}$$

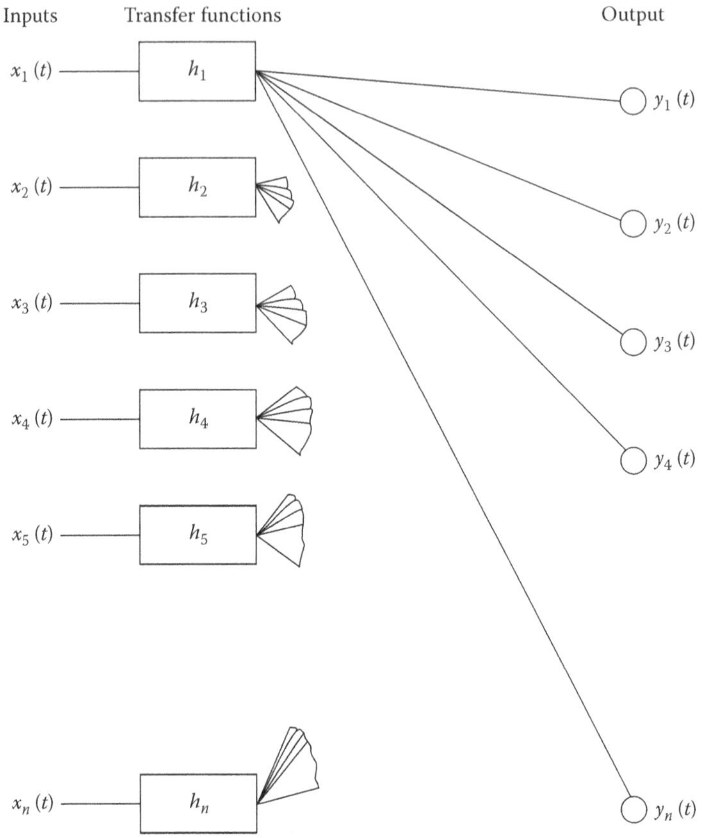

FIGURE 9.1
Block diagram for the system: $x_1(t)$, $x_2(t)$,…$x_n(t)$ = n inputs; h_1, h_2,…h_n = n transfer functions; $y_1(t)$, $y_2(t)$,…$y_m(t)$ = m outputs.

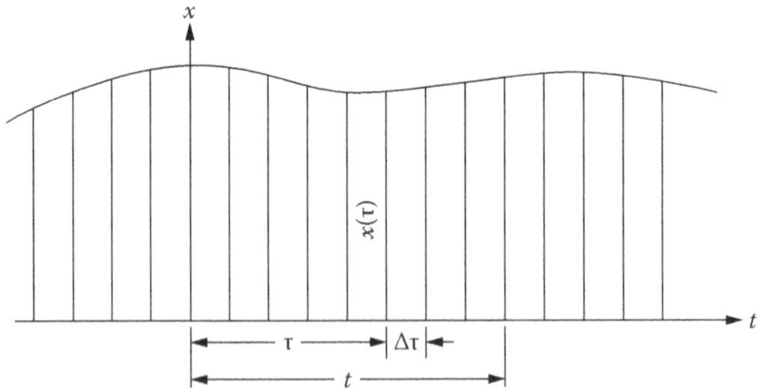

FIGURE 9.2
Input divided into infinitesimal impulses: t = time; τ = the value of time at which $x(\tau)$ is taken; $\Delta\tau$ = time width of impulse; $x(\tau)$ = value of the impulse at time τ.

9.2 Frequency Response Function

The frequency response function or the transfer function (or the system function, as it is sometimes known) is defined as the ratio of the complex output amplitude to the complex input amplitude for a steady-state sinusoidal input. (The frequency response function is the output per unit sinusoidal input at frequency ω.) Thus, the input is

$$x_i(t) = \bar{x}_i(\omega)e^{i\omega t} \tag{9.4}$$

and the corresponding output is

$$y_j = \bar{y}_j(\omega)e^{i\omega t} \tag{9.5}$$

where $\bar{x}_i(\omega)$ and $\bar{y}_j(\omega)$ are the complex amplitudes of the input and output, respectively. Then the frequency response function $H_{ij}(\omega)$ is

$$H_{ij}(\omega) = \frac{\bar{y}_j(\omega)}{\bar{x}_i(\omega)} \tag{9.6}$$

For sinusoidal input and output, Equation 9.6 becomes

$$\frac{\bar{y}_j(\omega)}{\bar{x}_i(\omega)} = \int_{-\infty}^{+\infty} h_{ij}(\tau)e^{-i\omega\tau}\, d\tau \tag{9.7}$$

It is therefore proven that the frequency response function is the Fourier transform of the unit impulse function.

9.3 Statistics of the Response

Since the linear process is assumed to be random, the results are based on statistical operations on the process. In this section, the pertinent statistical parameters will be derived. Referring back to Equation 9.3, we see that the total response y_j is the sum over all inputs. Thus,

$$y_j(t) = \sum_{i=1}^{n} \int_{-\infty}^{+\infty} h_{ij}(\tau)x_i(t-\tau)\, d\tau \tag{9.8}$$

The cross correlation between outputs $y_j(t)$ and $y_k(t)$ is defined as follows:

$$C_{jk}(\tau) = \lim_{T\to\infty}\frac{1}{2T}\int_{-T}^{+T} y_j(t)y_k(t+\tau)\,dt \tag{9.9}$$

From the definition of $C_{jk}(\tau)$ it is seen that

$$C_{kj}(\tau) = \langle y_k(t)y_j(t+\tau)\rangle$$

where

$$\langle(\,)\rangle = \lim_{T\to\infty}\frac{1}{2T}\int_{-T}^{+T}(\,)\mathrm{d}t$$

Substituting Equation 9.8 and rearranging,

$$\begin{aligned} C_{jk}(\tau) = \sum_{s=1}^{m}\sum_{r=1}^{n}\int_{-\infty}^{+\infty}du\int_{-\infty}^{+\infty}d\upsilon\{h_{js}(u)h_{kr}(\upsilon) \\ \times\left[\lim_{T\to\infty}\frac{1}{2T}\int_{-T}^{+T}x_s(t-u)x_r(t-\upsilon+\tau)\,dt\right]\right\} \end{aligned} \tag{9.10}$$

By definition of the cross correlation,

$$\begin{aligned} \lim_{T\to\infty}\frac{1}{2T}\int_{-T}^{+T}x_s(t-u)x_r(t-\upsilon+\tau)\,dt \\ = C_{rs}(u-\upsilon+\tau) \end{aligned} \tag{9.11}$$

Thus,

$$\begin{aligned} C_{jk}(\tau) = \sum_{s=1}^{n}\sum_{r=1}^{n}\int_{-\infty}^{+\infty}du\int_{-\infty}^{+\infty}d\upsilon \\ \times[h_{js}u)h_{kr}(\upsilon)C_{rs}(u-\upsilon+\tau)] \end{aligned} \tag{9.12}$$

The cross spectrum $G_{jk}(\omega)$ is defined as the Fourier transform of the cross correlation. The inverse Fourier transform relation is then

$$C_{jk}(\tau) = \frac{1}{2\pi}\int_{-\infty}^{+\infty} G_{jk}(\omega)e^{i\omega\tau}\, d\omega$$

Thus,

$$G_{jk}(\omega) = \int_{-\infty}^{+\infty} C_{jk}(\tau)e^{-i\omega\tau}d\tau \tag{9.13}$$

Note that

$$G_{kj}(\omega) = \int_{-\infty}^{+\infty} C_{kj}(\tau)e^{i\omega\tau}d\tau = \int_{-\infty}^{+\infty} C_{jk}(-\tau)e^{-i\omega\tau}d\tau$$

$$= \int_{-\infty}^{+\infty} C_{jk}(\theta)e^{i\omega\theta}d\theta = G_{jk}^{*}(\omega)$$

where G_{jk}^{*} is the complex conjugate of G_{jk}.

Substituting Equation 9.12, changing variables $\theta = u - v + \tau$, and using Equations 9.13 and 9.7,

$$G_{jk}(\omega) = \sum_{s=1}^{n}\sum_{r=1}^{n} H_{js}^{*}(\omega)H_{kr}(\omega)G_{rs}(\omega) \tag{9.14}$$

in which H_{js}^{*} is the complex conjugate of H_{js}. Equation 9.14 gives the cross spectrum of the outputs $G_{jk}(\omega)$ in terms of the cross spectrum of the inputs $G_{rs}(\omega)$. In matrix notation, Equation 9.14 can be written as

$$G^{o}(\omega) = H^{*}G^{i}H^{T} \tag{9.15}$$

where G^{o} is the output matrix of cross spectra, G^{i} is the input matrix of cross spectra, and H is the matrix of transfer functions. H^{T} denotes the transpose matrix of H, and H^{*} is the complex conjugate matrix of H. Thus,

$$G^{o}(\omega) = \begin{vmatrix} G_{11}^{o}(\omega) & G_{12}^{o}(\omega) & G_{13}^{o}(\omega) & \cdots & G_{1k}^{o}(\omega) \\ G_{21}^{o}(\omega) & & & & \\ \vdots & & & & \\ G_{j1}^{o}(\omega) & & \cdots & & G_{jj}^{o}(\omega) \end{vmatrix} \tag{9.16}$$

By virtue of the fact that $G_{ij}(\omega) = G_{ji}^*(\omega)$, the preceding matrix is a square Hermitian matrix. The input matrix G^{i} is

$$G^{\mathrm{i}}(\omega) = \begin{vmatrix} G^{\mathrm{i}}_{11}(\omega) & G^{\mathrm{i}}_{12}(\omega) & G^{\mathrm{i}}_{13}(\omega) & \cdots & G^{\mathrm{i}}_{15}(\omega) \\ G^{\mathrm{i}}_{21}(\omega) & & & & \\ \vdots & & & & \\ G^{\mathrm{i}}_{r1}(\omega) & & \cdots & & G^{\mathrm{i}}_{rr}(\omega) \end{vmatrix} \tag{9.17}$$

G^{i} is also Hermitian of order $r \times r = s \times s = r \times s$. The transfer function matrix is a $k \times r$ complex matrix (not Hermitian):

$$H(\omega) = \begin{vmatrix} H_{11}(\omega) & H_{12}(\omega) & \cdots & H_{1r}(\omega) \\ H_{21}(\omega) & & & \\ \vdots & & & \\ H_{k1}(\omega) & & \cdots & H_{kr}(\omega) \end{vmatrix} \tag{9.18}$$

9.4 Important Quantities Derivable from the Cross Spectrum

The cross spectrum can be used as a starting point to derive several important quantities. The spectrum of the response at point j is obtained by letting $k = j$ in Equation 9.14. The autocorrelation is obtained by letting $k = j$ in Equation 9.9. The mean-square response is further obtained from the autocorrelation by letting $\tau = 0$. If the Fourier inverse of Equation 9.13 is used to determine mean square, then

$$\begin{aligned} C_{jj}(0) &= \frac{1}{2\pi}\int_{-\infty}^{+\infty} G_{jj}(\omega)\, d\omega \\ &= \text{Mean Square} = M_j^2 \\ &= \lim_{T\to\infty}\frac{1}{2T}\int_{-T}^{+T} y_j^2(t)\, dt \end{aligned} \tag{9.19}$$

If the mean square is desired over a frequency band $\Delta\Omega = \Omega_2 - \Omega_1$, then it is given by

$$(M_j^2)_{\Delta\Omega} = \frac{1}{2\pi}\int_{\Omega_1}^{\Omega_2} G_{jj}(\omega)\, d\omega \tag{9.20}$$

The mean value of $y_j(t)$ is defined as

$$\bar{M}_j = \lim_{T\to\infty} \frac{1}{2T} \int_{-T}^{+T} y_j(t)\, dt \tag{9.21}$$

The variance σ_j^2 is defined as the mean square value about the mean

$$\sigma_j^2 = \lim_{T\to\infty} \frac{1}{2\pi} \int_{-T}^{+T} [y_j(t) - \bar{M}_j]^2\, dt \tag{9.22}$$

The square root of the variance is known as the standard deviation. By using Equations 9.19 and 9.21, Equation 9.22 can be written as

$$\sigma_j^2 = M_j^2 - (\bar{M}_j)^2 \tag{9.23}$$

The mean, the variance, and the standard deviation are three important parameters involved in probability distributions. Note that if the process is one with zero mean value, then the variance is equal to the mean square, and the standard deviation is the root mean square.

The preceding quantities are associated with the ordinary spectrum rather than the cross spectrum. An important physical quantity associated with the cross spectrum is the coherence, which is defined as

$$\gamma_{jk}^2(\omega) = \frac{|G_{jk}(\omega)|^2}{G_{jj}(\omega) G_{kk}(\omega)} \tag{9.24}$$

The lower limit of γ_{jk}^2 must be zero since the lower limit of $G_{jk}(\omega)$ is zero. This corresponds to no correlation between the signals at j and k. In addition, $\gamma_{jk}^2 \le 1$. Going back to Equation 9.24, we see that if there is only one input, then

$$G_{jk}(\omega) = H_{js}^* H_{kr} G_{rr} \tag{9.25}$$

Thus,

$$\gamma_{jk}^2 = \frac{H_{js}^* H_{kr} H_{js} H_{kr}^* G_{rr}^2}{H_{js}^* H_{js} G_{rr} H_{kr}^* H_{kr} G_{rr}} = 1 \tag{9.26}$$

Therefore, the field is completely coherent for a single input to the system.

In an acoustic field, sound emanating from a single source is coherent. If the coherence is less than unity, then the field is partially coherent. The partial coherence effect is sometimes due to the fact that the source is of finite extent. It is also sometimes due to the fact that several sources are causing the radiation and these sources are correlated in some way with each other.

9.5 The Cross Spectrum in Terms of Fourier Transforms

The cross spectrum can also be expressed in terms of Fourier transforms alone. To see this, start with the basic definition of cross spectrum as given by Equation 9.23, where

$$C_{jk}(\tau) = \lim_{T\to\infty}\frac{1}{2T}\int_{-T}^{+T} y_j(t)y_k(t+\tau)\,dt \tag{9.27}$$

Thus,

$$G_{jk}(\omega) = \int_{-\infty}^{+\infty}\left\{\lim_{T\to\infty}\frac{1}{2T} \times\int_{-T}^{+T} y_j(t)y_k(t+\tau)\,dt\right\}e^{-i\omega\tau}d\tau \tag{9.28}$$

Letting $t = u$ and $t + \tau = \upsilon$, we have

$$G_{jk} = \int_{-\infty}^{+\infty}\left\{\lim_{T\to\infty}\frac{1}{2T} \times\int_{-T}^{+T} y_j(u)y_k(\upsilon)\,du\right\}e^{-i\omega(\upsilon-u)}d\upsilon \tag{9.29}$$

$$= \int_{-\infty}^{+\infty}\left\{\lim_{T\to\infty}\frac{1}{2T} \times\int_{-T}^{+T} y_j(u)e^{i\omega u}\,du\right\}y_k(\upsilon)e^{-i\omega\upsilon}d\upsilon \tag{9.30}$$

The next step is true only under the condition that the process is ergodic. In this case, the last equation can be written as

$$G_{jk}(\omega) = \lim_{T\to\infty} \frac{1}{2T}\left(\int_{-T}^{+T} y_j(u)e^{i\omega u}\, du\right) \times \left(\int_{-T}^{+T} y_k(\upsilon)e^{-i\omega_\upsilon} d\upsilon\right) \tag{9.31}$$

This last relation can then be written as

$$G_{jk}(\omega) = \lim_{T\to\infty} \frac{\bar{y}_j^*(T,\omega)\bar{y}_k(T,\omega)}{2T} \tag{9.32}$$

where

$$\bar{y}_j(T,\omega) = \int_{-T}^{+T} y_j(t)e^{-i\omega t}\, dt \tag{9.33}$$

and y_j^* is the complex conjugate of y_j.

$$\bar{y}_k(T,\omega) = \int_{-T}^{+T} y_k(t)e^{-i\omega t}\, dt \tag{9.34}$$

Equation 9.32 expresses the cross spectrum in terms of the limit of the product of truncated Fourier transforms.

9.6 The Conceptual Meaning of Cross Correlation, Cross Spectrum, and Coherence

Given two functions of time $x(t)$ and $y(t)$, the cross correlation between these two functions is defined mathematically by the following formula:

$$C(x,y,\tau) = \lim_{T\to\infty} \frac{1}{2T} \int_{-T}^{+T} x(t)y(t+\tau)\, dt \tag{9.35}$$

This formula states that we take x at any time t, multiply it by y at a time $t + \tau$ (i.e., at a time τ later than t), and sum the product over all values ($-T < t < +T$). The result is then divided by $2T$. In real systems, naturally, T is finite, and the meaning of ∞ in the formula is that various values of T must be tried to make sure that the same answer results independently of T.

For two arbitrary functions of time, the formula in Equation 9.35 has no real meaning. It is only when the two signals have something to do with each other that the cross correlation tells us something. To see this point clearly, consider an arbitrary random wave train moving in space. (It could be an acoustic wave, an elastic wave, an electromagnetic wave, etc.) Let $x(t) = x_1(t)$ be the response (pressure, stress, etc.) at one point, and $y(t) = x_2(t)$ be the response at another point. Now, form the cross correlation between x_1 and x_2 (the limit is eliminated because it is understood):

$$C(x_1, y_2, \tau) = \frac{1}{2T} \int_{-T}^{+T} x_1(t) x_2(t+\tau)\, dt \tag{9.36}$$

When the points coincide (i.e., $x_1 = x_2$), the relation becomes

$$C(x_1, \tau) = \frac{1}{2T} \int_{-T}^{+T} x_1(t) x_1(t+\tau)\, dt \tag{9.37}$$

and, if $\tau = 0$,

$$C(x_1, 0) = \frac{1}{2T} \int_{-T}^{+T} x_1^2(t)\, dt \tag{9.38}$$

which is, by definition, the mean square value of the response at point x_1.

For other values of τ, Equation 9.37 defines the autocorrelation at point 1. It is the mean value between the response at one time and the response at another time τ later than t. Thus, Equation 9.36 is the mean product between the response at point 1 and the response at point 2 at a time τ later.

Now, going back to the random wave train, let us assume that it is traveling in a nondispersive medium (i.e., with velocity independent of frequency). It is seen that if the wave train leaves point 1 at time t (Figure 9.3) and travels through the system with no distortion, then

$$y(t) = x_2(t) = A x_1(t - \tau_1) \tag{9.39}$$

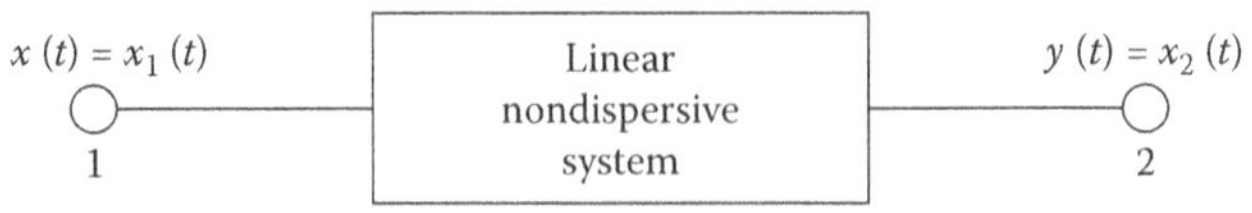

FIGURE 9.3
Input and output in a linear system: $x(t)$ = input; $y(t)$ = output.

where A is some decay constant giving the amount that the wave has decreased in amplitude from point 1 to point 2, and τ is the time of travel from 1 to 2. Forming the cross correlation $C(x_1, x_2, \tau)$ gives

$$\begin{aligned} C(x_1,x_2,\tau) &= \frac{1}{2T}\int_{-T}^{+T} x_1(t)Ax_1(t+\tau-\tau_1)\,dt \\ &= AC(x_1,x_2,\tau-\tau_1) \end{aligned} \tag{9.40}$$

Thus, the cross correlation of a random wave train in a nondispersive system is exactly the same form as the autocorrelation of the wave at the starting point, except that the peak occurs at a time delay corresponding to the time necessary for the wave to travel between the points. In the absence of attenuation, the wave is transmitted undisturbed in the medium. However, in most cases, it is probable that the peak is attenuated somewhat as the wave progresses. It is thus seen that cross correlation is an extremely useful concept for measuring the time delay of a propagating random signal. In the preceding case, it had to be assumed that the signal was propagating in a nondispersive medium and that, when the cross correlation was done, the signal was actually being traced as it moved through the system.

Consider the meaning of cross correlation if the system was dispersive (i.e., if the velocity was a function of frequency). White[4] has addressed this question and has demonstrated that time delays in the cross correlation can still be measured with confidence if the signal that is traveling is band-limited noise. For dispersive systems where the velocity is a function of frequency, it has been pointed out in the literature that time delays can also be obtained. For this case, the following cross spectrum is formed:

$$S_{12}(\omega) = \int_{-\infty}^{+\infty} C_{12}(\tau)e^{-i\omega\tau}d\tau \tag{9.41}$$

The cross spectrum is a complex number and can be written in terms of amplitude and phase angle $\theta_{12}(\omega)$ as follows:

$$S_{12}(\omega) = |S_{12}(\omega)|e^{-i\theta_{12}(\omega)} \tag{9.42}$$

The phase angle $\theta_{12}(\omega)$ is actually the phase between input and output at frequency ω. The time delay from input to output is then

$$\tau(\omega) = \theta_{12}(\omega)/\omega \tag{9.43}$$

Suppose that the signal has lost its propagating properties in that it has reflected many times and set up a reverberant field in the system. Consider the physical meaning of cross correlation in this case. To answer this question partially, examine an optical field. In optical systems, extensive use has been made of the concept of partial coherence.

At the beginning of this section, two functions, $x(t)$ and $y(t)$, were chosen, and the cross correlation between them was formed. It was pointed out that if the two functions are completely arbitrary, then there is no real physical meaning to the cross correlation. However, if the two functions are descriptions of response at points of a field, then there is a common ground to interpret cross correlation. Thus, the cross correlation and any function that is derived from it give some measure of the dependence of the vibrations at one point on the vibrations at the other point. This is a general statement, and to tie it down the concept of coherence has been used.

In the optical case, suppose that light is coming into two points in a field. If the light comes from a small, single source of narrow spectral range (i.e., almost single frequency), then the light at the two field points is dependent. If each point in the field receives light from a different source, then the light at the two field points is independent. The first case is termed coherent, and the field at the points is highly correlated (or dependent). The second case is termed incoherent, and the field between the two points is uncorrelated (independent).

These are the two extreme cases, and between them there are degrees of coherence (i.e., partial coherence). Just as in everyday usage, coherence is analogous to clarity or sharpness of the field, whereas incoherence is tantamount to haziness or "jumbledness." The same idea is used when speaking about someone's speech or a written article. If it is concise and presented clearly, it is coherent. If the ideas and presentation are jumbled, it can be called incoherent.

Single-frequency radiation is coherent radiation; radiation with a finite bandwidth is not. The partial coherence associated with finite spectral bandwidth is called the temporal (or time-wise) coherence. On the other hand, light or sound emanating from a single source gives coherent radiation, but a point source is never actually obtained. The partial coherence effect, due to the fact that the source is of finite extent, is termed space coherence. The point source gives unit coherence in a system, whereas an extended source gives coherence somewhat less than unity.

The square of coherence $Y_{12}(\omega)$ between signals at points 1 and 2 at frequency ω is defined as

$$\gamma_{12}^2(\omega) = \frac{|S_{12}(\omega)|^2}{S_{11}(\omega)S_{22}(\omega)} \tag{9.44}$$

where

$$S_{12}(\omega) = \int_{-\infty}^{+\infty} C_{12}(\tau)e^{-i\omega\tau}\, d\tau \tag{9.45}$$

in which $C_{12}(\tau)$ is the cross correlation between signals at points 1 and 2. The function $S_{12}(\omega)$ is the cross spectrum between signals at points 1 and 2, and $S_{11}(\omega)$, $S_{22}(\omega)$ are the autospectra of the signals at points 1 and 2, respectively. Wolf[7] has other ways of defining coherence by functions called complex degree of coherence or mutual coherence function, but it all amounts conceptually to the same cross spectrum as given by Equation 9.44.

Although there are formal proofs that $\gamma_{12}(\omega)$ is always between 0 and 1, one can reason this out nonrigorously by going back to the basic physical ideas associated with correlation and coherence. If the signals at two points are uncorrelated and therefore incoherent, the cross correlation is zero and thereby $\gamma_{12}(\omega)$ is zero. If the signals are perfectly correlated, then this is tantamount to saying that the signals in the field are a result of input to the system from a single source, as shown in Figure 9.4.

As seen before, the cross spectrum $S_{yz}(\omega)$ can be written in terms of the input spectrum $S_x(\omega)$ and the transfer functions $Y_z(i\omega)$ and $Y_y(i\omega)$ as follows:

$$S_{yz}(i\omega) = Y_y(i\omega)Y_z^*(i\omega)S_x(\omega) \tag{9.46}$$

Thus,

$$\gamma_{yz}^2(\omega) = \frac{\left|S_{yz}(\omega)\right|^2}{S_{yy}(\omega)S_{zz}(\omega)} \tag{9.47}$$

$$= \frac{Y_y(i\omega)Y_z^*(i\omega)Y_y^*(i\omega)Y_z(i\omega)S_x^2(\omega)}{\left|Y_y(i\omega)\right|^2 S_x(\omega)\left|Y_z(i\omega)\right|^2 S_x(\omega)} = 1 \tag{9.48}$$

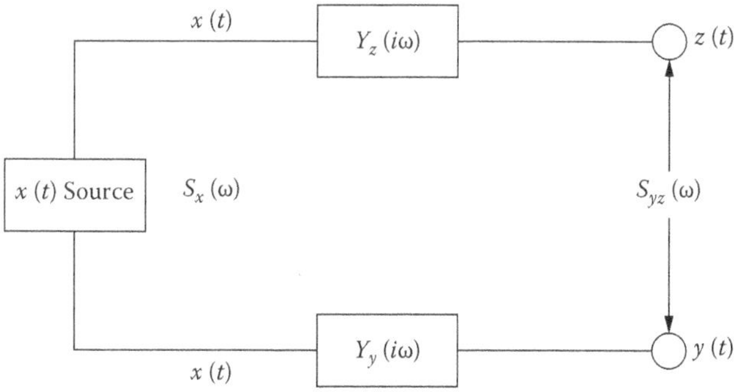

FIGURE 9.4
Single-input or coherent system: $x(t)$ = input; $S_x(\omega)$ = spectrum of input; $Y_z(i\omega)$ = transfer function for z output; $Y_y(i\omega)$ = transfer function for y output; $z(t)$, $y(t)$ = outputs.

$$\text{So that } 0 \leq \gamma_{12}(\omega) \leq 1 \text{ (complete coherence)} \tag{9.49}$$

Between the cases of complete coherence and complete incoherence, there are many degrees of partial coherence.

References

1. Middleton, D. 1960. *An introduction to statistical communication theory.* New York: McGraw–Hill Book Co.
2. Bendat, J. S., and Piersol, A. G. 1966. *Measurement and analysis of random data.* New York: John Wiley & Sons, Inc.
3. Lee, Y. W. 1960. *Statistical theory of communication.* New York: John Wiley & Sons, Inc.
4. White, P. H. 1966. Cross correlation in structural vibration. Final report for USN Marine Engineering Laboratory under contract N600(61533)63705. Measurement Analysis Corp.
5. Born, M., and Wolf, E. 1959. *Principles of optics,* 14. New York: MacMillan Co.
6. Beran, M. J., and Parrent, G. B., Jr. 1964. *Theory of partial coherence.* Englewood Cliffs, NJ: Prentice Hall, Inc.
7. Wolf, E. 1955. *Proceedings of Royal Society (London)* A230:246.
8. Greenspon, J. E. 1992. Acoustics, linear. In *Encyclopedia of physical science and technology,* vol. 1, 114–152. New York: Academic Press, Inc.–Elsevier.

10

*Statistical Acoustics**

10.1 Physical Concept of Transfer Function

In Chapter 2 it was shown that H_{ij} was the transfer function that gave the output at j per unit sinusoidal input at i. Suppose there is an acoustic field generated by a group of sound sources and that these sources are surrounded by an imaginary surface S_o, as shown in Figure 10.1. Through each element of S_o sound passes into the field. Thus, each element of S_o, denoted by ds, can be considered a source that radiates sound into the field. Consider the pressure $dp(P,\omega)$ at field point P at frequency ω due to radiation out of element ds:

$$dp(P,\omega) = H_p(P,S,\omega)p(S,\omega)\,ds \tag{10.1}$$

where $H_p(P, S, \omega)$ is the pressure at field point P per unit area of S due to a unit sinusoidal input pressure on S. The total pressure in the field at point P is

$$p(P,\omega) = \int_{S_o} H_p(P,S,\omega)p(S,\omega)\,ds \tag{10.2}$$

If motion (e.g., acceleration) of the surface S is considered instead of pressure, the counterpart to Equation 10.2 is

$$p(P,\omega) = \int_{S_o} H_a(P,S,\omega)a(S,\omega)\,ds \tag{10.3}$$

where $H(P, S, \omega)$ is the transfer function associated with acceleration; that is, it is the pressure at field point P per unit area of S due to a unit sinusoidal input acceleration of S.

Applying these ideas to Equation 9.15, it is seen that the cross spectrum of the field pressure can immediately be written in terms of the cross spectrum

* This chapter was published in *Encyclopedia of Physical Science and Technology*, vol. 1, Greenspon, J.E., 114–152. Copyright Elsevier (1992).

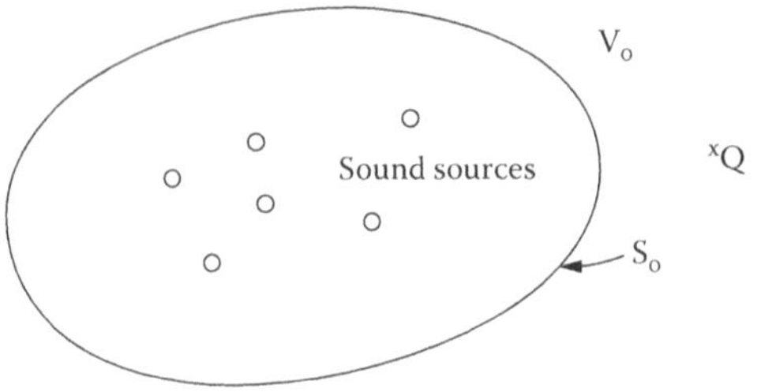

FIGURE 10.1
Surface surrounding the sources. S_o = surrounding surface; V_o = volume outside the surface; P, Q = field points.

of the surface pressure or the cross spectrum of the surface acceleration. For surface pressure, Equation 9.15 becomes

$$G(P,Q,\omega) = \int_{S_i}\int_{S_r} H_p^*(P,S_i,\omega)H_p(Q,S_r,\omega) \times G(S_i,S_r,\omega)\, ds_i\, ds_r \tag{10.4}$$

Comparing Equation 10.4 with Equation 9.15, we find that the points j, k become field points P, Q. $G(P, Q, \omega)$ is the cross spectrum of pressure at field points P, Q. The transfer functions $H_{js}^*(\omega)$ become $H_p^*(P, S_i, \omega)$, in which S_i is the surface point or ith input point. $H_{kr}(\omega)$ becomes $H_p(Q, S_r, \omega)$, where S_r is the other surface point or rth input point. $G(S_i, S_r, \omega)$ is the input cross spectrum, which is the cross spectrum of surface pressure. The summations over r and s become integrals over S_i and S_r. For acceleration

$$G(P,Q,\omega) = \int_{S_i}\int_{S_r} H_a^*(P,S_i,\omega)H_a(Q,S_r,\omega) \times A(S_i,S_r,\omega)\, ds_i\, ds_r \tag{10.5}$$

The transfer functions are those for acceleration, and $A(S_i, S_r, \omega)$ is the cross spectrum of the surface acceleration. The relation can be written for any other surface input, such as velocity.

10.2 Response in Terms of Green's Functions

The Green's functions for single-frequency systems were taken up in a previous section. The transfer function for pressure is associated with the Green's function that vanishes over the surface. Thus,

$$H_p(P,S,\omega) = \frac{\partial g_1(P,S,\omega)}{\partial n} \tag{10.6}$$

The transfer function for acceleration is associated with the Green's function, whose normal derivative vanishes over S_o. Thus,

$$H_a(P,S,\omega) = \rho g_2(P,S,\omega)$$

The statistical relations for the field pressure can therefore immediately be written in terms of the cross spectrum of pressure or acceleration over the surface surrounding the sources:

$$\begin{aligned} G(P,Q,\omega) = \int_{S_i}\int_{S_r} & \frac{\partial g_1^*(P,S_i,\omega)}{\partial n_i}\frac{\partial g_1(Q,S_r,\omega)}{\partial n_r} \\ & \times G(S_i,S_r,\omega)\, ds_i\, ds_r \end{aligned} \tag{10.7}$$

or

$$\begin{aligned} G(P,Q,\omega) = \int_{S_i}\int_{S_r} & \rho^2 g_2^*(P,S_i,\omega) g_2(Q,S_r,\omega) \\ & \times A(S_i,S_r,\omega)\, ds_i\, ds_r \end{aligned} \tag{10.8}$$

These relations give the cross spectrum of the field pressure as a function of either the cross spectrum of the surface pressure $G(S_i, S_r, \omega)$ or the cross spectrum of the surface acceleration $A(S_i, S_r, \omega)$. Equation 10.7 was derived by Parrent[6] using a different approach. At this point, one should review the relationship between Equations 10.7 and 10.8 for acoustic systems and Equation 9.14 for general linear systems. It is evident that the inputs to Equations 10.7 and 10.8 are $G(S_i, S_r, \omega)$ and $A(S_i, S_r, \omega)$, respectively, and the output is $G(P, Q, \omega)$ in both cases. The frequency response functions are the transfer functions described by Equation 10.6.

10.3 Statistical Differential Equations Governing the Sound Field

Consider the general case where there are source terms present in the field equation. This is tantamount to saying that, outside the series of main radiating sources that have been surrounded by a surface (see Figure 10.2), there are other sources arbitrarily located in the field. For example, in the case of

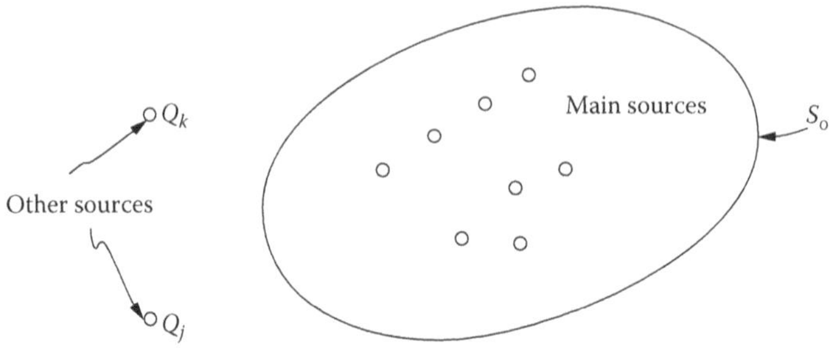

FIGURE 10.2
Surface surrounding main sources with presence of other field sources. S_o = surrounding surface; Q_k, Q_j = strengths of other sources not within S_o.

turbulence surrounding a moving structure, the turbulent volume constitutes such a source, whereas the surface of the structure surrounds all the other vibrating sources. The equation governing the propagation of sound waves in the medium is

$$\nabla^2 p(P,t) - \frac{1}{c_0^2}\frac{\partial^2 p(P,t)}{\partial t^2} = V(Q,t) \tag{10.9}$$

where

$$V(Q,t) = \sum_i V_i(Q_i,t) \tag{10.10}$$

In the preceding equation, V is a general source term that may consist of a series of sources at various points in the medium. Actually, the medium being considered is bounded internally by S_o, so the sources inside S_o are not in the medium. The sources at Q_i, however, are in the medium.

The various types of source terms that can enter acoustical fields arise from the injection of mass, momentum, heat energy, or vorticity into the field. These are discussed by Morse and Ingard[7] and will not be treated here. It is assumed that the source term $V(Q, t)$ is a known function of space and time or, if it is a random function, that some statistical information is known about it, such as its cross correlation or cross spectrum.

In cases where the field is random, a statistical description has to be used. The cross-correlation function $\Gamma(P_1, P_2, \tau)$ between pressures at field points P_1 and P_2 is defined as

$$\Gamma(P_1,P_2,\tau) = \lim_{T\to\infty}\frac{1}{2T} \times \int_{-T}^{+T} p(P_1,t)p(P_2,t+\tau)dt \tag{10.11}$$

(To the author's knowledge, one of the first pieces of work on correlation in wave fields was the paper of Marsh.[8])

The Fourier transform $U(P, \omega)$ of the pressure $p(P, t)$ is

$$U(P,\omega) = \int_{-\infty}^{+\infty} p(P,t)e^{-i\omega t}dt \tag{10.12}$$

Taking the inverse, we can write the preceding equation as

$$\nabla^2 p = \frac{1}{2\pi}\int_{-\infty}^{+\infty} \nabla^2 U(P,\omega)e^{i\omega t}d\omega \tag{10.13}$$

and also from the Fourier transform of $V(Q, t)$—that is,

$$W(Q,\omega) = \int_{-\infty}^{+\infty} V(Q,t)e^{-i\omega t}dt \tag{10.14}$$

and its inverse

$$V(Q,t) = \frac{1}{2\pi}\int_{-\infty}^{+\infty} W(Q,\omega)e^{i\omega t}d\omega \tag{10.15}$$

Substitution into the original nonhomogeneous wave equation (Equation 10.9) gives

$$\frac{1}{2\pi}\int_{-\infty}^{+\infty}\left[\nabla^2 U + \frac{\omega^2}{c_0^2}U - W\right]d\omega = 0 \tag{10.16}$$

Thus, for this relation to hold for all P and all ω, there must be

$$\nabla^2 U(P,\omega) + k^2 U(P,\omega) = W(Q,\omega) \tag{10.17}$$

The cross spectrum $G(P_1, P_2, \omega)$ between the pressures at P_1 and P_2 at frequency ω is defined in terms of the cross correlation $\Gamma(P_1, P_2, \tau)$:

$$G(P_1,P_2,\omega) = \int_{-\infty}^{+\infty} \Gamma(P_1,P_2,\tau)e^{i\omega\tau}d\tau \tag{10.18}$$

and the inverse is

$$\Gamma(P_1,P_2,\tau) = \frac{1}{2\pi}\int_{-\infty}^{+\infty} G(P_1,P_2,\omega)e^{-i\omega\tau}d\omega \tag{10.19}$$

Thus, $\Gamma(P_1, P_2, \tau)$ can be written as

$$\Gamma(P_1,P_2,\tau) = \lim_{T\to\infty}\frac{1}{2T} \times \int_{-T}^{+T}\left[\frac{1}{2\pi}\int_{-\infty}^{+\infty} U(P_1,\omega)e^{-i\omega t}d\omega\right] \times\left[\frac{1}{2\pi}\int_{-\infty}^{+\infty} U(P_2,\omega)e^{-i\omega(t+\tau)}d\omega\right]dt \tag{10.20}$$

Thus,

$$\nabla_2^2\Gamma(P_1,P_2,\tau) = \lim_{T\to\infty}\frac{1}{2T} \times \int_{-T}^{+T}\left[\frac{1}{2\pi}\int_{-\infty}^{+\infty} U(P_1,\omega)e^{-i\omega t}d\omega\right] \times\left[\frac{1}{2\pi}\int_{-\infty}^{+\infty} \nabla_2^2 U(P_2,\omega)e^{-i\omega(t+\tau)}d\omega\right]dt \tag{10.21}$$

where $\nabla^2{}_2$ stands for operations performed on the $P_2(x_2, y_2, z_2)$ coordinates. Also,

$$\frac{\partial^2\Gamma(P_1,P_2,\tau)}{\partial\tau^2} = \lim_{T\to\infty}\frac{1}{2T} \times \int_{-T}^{+T}\left[\frac{1}{2\pi}\int_{-\infty}^{+\infty} U(P_1,\omega)e^{-i\omega t}\, d\omega\right] \times\left[\frac{1}{2}\int_{-\infty}^{+\infty} (-\omega^2)U(P_2,\omega)e^{-i\omega(t+\tau)}d\omega\right]dt \tag{10.22}$$

Thus,

$$\nabla_2^2\Gamma(P_1,P_2,\tau) - \frac{1}{c_0^2}\frac{\partial^2\Gamma(P_1,P_2,\tau)}{\partial\tau^2} = \langle p(P_1,t)V(Q_2,t+\tau)\rangle \tag{10.23}$$

It should be clear from an analysis similar to that given previously that the following relation also holds:

$$\nabla_1^2 \Gamma(P_2, P_1, \tau) - \frac{1}{c_0^2} \frac{\partial^2 \Gamma(P_2, P_1, \tau)}{\partial \tau^2} = \langle p(P_2, t) V(Q_1, t+\tau) \rangle \tag{10.24}$$

Equations 10.23 and 10.24 are extensions of the equation obtained by Eckart.[9] If the source term were zero, then

$$\begin{aligned} \nabla_2^2 \Gamma(P_1, P_2, \tau) - \frac{1}{c_0^2} \frac{\partial^2 \Gamma(P_1, P_2, \tau)}{\partial \tau^2} = 0 \\ \nabla_1^2 \Gamma(P_2, P_1, \tau) - \frac{1}{c_0^2} \frac{\partial^2 \Gamma(P_2, P_1, \tau)}{\partial \tau^2} = 0 \end{aligned} \tag{10.25}$$

However, since

$$\Gamma(P_2, P_1, \tau) = \Gamma(P_1, P_2, -\tau) \tag{10.26}$$

and

$$\frac{\partial^2}{\partial \tau^2} = \frac{\partial^2}{\partial(-\tau)^2} \tag{10.27}$$

then

$$\nabla_{1,2}^2 \Gamma(P_1, P_2, \tau) - \frac{1}{c_0^2} \frac{\partial^2 \Gamma(P_1, P_2, \tau)}{\partial \tau^2} = 0 \tag{10.28}$$

From the preceding relations, it is seen that the cross correlation is propagated in the same way that the original pressure wave propagates, except that real time t is replaced by correlation time r.

The nonhomogeneous counterparts given by Equations 10.23 and 10.24 state that the source term takes the statistical form of the cross correlation between the pressure p at a reference point and the source function V. Taking the Fourier transform of Equations 10.23 and 10.24, we see that the cross spectrum satisfies

$$\begin{aligned}
&\nabla_2^2 G(P_1,P_2,\omega)+k^2 G(P_1,P_2,\omega)\\
&\quad=\Phi_2(P_1,Q_2,\omega)\\
&\nabla_1^2 G(P_2,P_1,\omega)+k^2 G(P_2,P_1,\omega)\\
&\quad=\Phi_1(P_2,Q_1,\omega)
\end{aligned} \tag{10.29}$$

where Φ_1 and Φ_2 are the Fourier transforms of the cross correlation between the reference pressure and source function—that is,

$$\begin{aligned}
&\Phi_2(P_1,Q_2,\omega)\\
&\quad=\int_{-\infty}^{+\infty}\langle p(P_1,t)V(Q_2,t+\tau)\rangle e^{-i\omega\tau}d\tau\\
&\Phi_1(P_2,Q_1,\omega)\\
&\quad=\int_{-\infty}^{+\infty}\langle p(P_2,t)V(Q_1,t+\tau)\rangle e^{-i\omega\tau}d\tau
\end{aligned} \tag{10.30}$$

Thus, Φ_1 and Φ_2 are cross-spectrum functions between the pressure and source term.

In Equations 10.23, 10.24, 10.29, and 10.30, it is important to note that one point is being used as a reference point and the other is the actual variable. For example, in Equation 10.24, the varying is being done in the $P_1(x_1,y_1,z_1)$ coordinates; thus, all cross correlations are performed with P_2 fixed. Conversely, in Equation 10.23, all the operations are being carried out in the P_2 space, with P_1 remaining fixed.

References

1. Lamb, H. 1945. *Hydrodynamics.* New York: Dover Publications.
2. Rayleigh, J. W. S. 1945. *The theory of sound.* New York: Dover Publications.
3. Rayleigh, J. W. S. 1945. *The theory of sound,* 143. New York: Dover Publications.
4. Sommerfeld, A. 1964. *Partial differential equations in physics,* 189. New York: Academic Press.
5. Morse, P. M. 1948. *Vibration and sound,* 295. New York: McGraw–Hill Book Co.
6. Parrent, G. B., Jr. 1959. On the propagation of mutual coherence. *Journal of Optical Society of America* 49:787.
7. Morse, P. M., and Ingard, K. U. 1961. Linear acoustic theory. In *Encyclopedia of physics,* vol. XI/1, Acoustics I, 26. New York: Springer–Verlag.

8. Marsh, H. W. 1951. Correlation in wave fields. USN Underwater Sound Lab. tech. mem. 921-54-51.
9. Eckart, C. 1953. The theory of noise in continuous media. *Journal of the Acoustical Society of America* 25:195.
10. Greenspon, J. E. 1992. Acoustics, linear. In *Encyclopedia of physical science and technology*, vol. 1, 114–152. New York: Academic Press, Inc.–Elsevier.

11

*Statistics of Structures**

11.1 Integral Relation for the Response

Let the loading (per unit area) on the structure be represented by the function $f(r_0, t)$, where r_0 is the position vector of a loaded point on the body with respect to a fixed system of axes, as shown in Figure 11.1. Let the unit impulse response be $h(r, r_0, t - \theta)$; this is the output at r corresponding to a unit impulse at $t = 0$ and at location r_0.

The response at r at time t due to an arbitrary distributed excitation $f(r_0, t)$ can then be written:

$$w(\mathbf{r}, t) = \int_{\mathbf{r}_0} d\mathbf{r}_0 \int_{-\infty}^{t} f(\mathbf{r}_0, \theta) h(\mathbf{r}, \mathbf{r}_0, t - \theta) d\theta \tag{11.1}$$

The integration is taken over the whole loaded surface denoted by r_0. Since the loading is usually random in nature, only the statistics of the response—that is, the mean square values, the power spectral density, and so on—are determinable. Thus, let $U = t - \theta$ and form the cross correlation of the response at two points r_1 and r_2. This cross correlation is denoted by R_w and is

$$R_w(\mathbf{r}_1, \mathbf{r}_2, \tau) = \lim_{T\to\infty} \frac{1}{2T} \int_{-T}^{+T} \left[\int_{\mathbf{r}_0} d\mathbf{r}_0 \int_{-\infty}^{+\infty} f(\mathbf{r}_0, t - U_1) h(\mathbf{r}_1, \mathbf{r}_0, U_1) dU_1 \right] \left[\int_{\mathbf{r}_0} d\mathbf{r}'_0 \int_{-\infty}^{+\infty} f(\mathbf{r}'_0, t - U_2 + \tau) h(\mathbf{r}_2, \mathbf{r}'_0 U_2)\, dU_2 \right] dt \tag{11.2}$$

* This chapter was published in *Encyclopedia of Physical Science and Technology*, vol. 1, Greenspon, J.E., 114–152. Copyright Elsevier (1992).

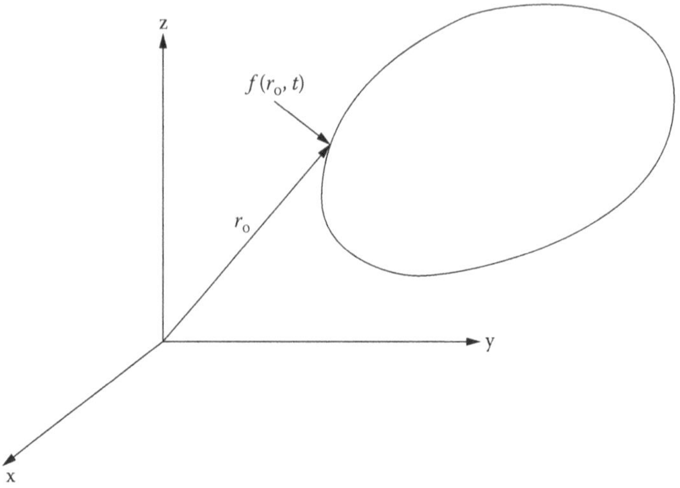

FIGURE 11.1
The loaded structure $f(r_o, t)$ = force component at location r_o and time t; $w(r, t)$ = deflection component at r at time t.

or

$$R_w(\mathbf{r}_1, \mathbf{r}_2, \tau) = \int_{\mathbf{r}_0}\int_{\mathbf{r}_0'} d\mathbf{r}_0 d\mathbf{r}_0' \int_{-\infty}^{+\infty}\int_{-\infty}^{+\infty} h(\mathbf{r}_1, \mathbf{r}_0', U_1)\, h(\mathbf{r}_2, \mathbf{r}_0', U_2) dU_1\, dU_2$$
$$\times\left[\lim_{T\to\infty}\frac{1}{2T}\int_{-T}^{+T} f(\mathbf{r}_0, t-U_1) f(\mathbf{r}_0', t-U_2+\tau)dt\right. \tag{11.3}$$

We assume a stationary process so that the loading is only a function of the difference of the times $t - U_2 + \tau$ and $t - U_1$. Let

$$\begin{aligned}\tau_3 &= (t-U_2+\tau)-(t-U_1)\\ &= U_1 - U_2 + \tau\end{aligned} \tag{11.4}$$

Then,

$$\begin{aligned}R_w(\mathbf{r}_1, \mathbf{r}_2, \tau) = \int_{\mathbf{r}_0}\int_{\mathbf{r}_0'} d\mathbf{r}_0 d\mathbf{r}_0' &\int_{-\infty}^{+\infty}\int_{-\infty}^{+\infty}\\ &h(\mathbf{r}_1, \mathbf{r}_0, U_1) h(\mathbf{r}_2, \mathbf{r}_0', U_2)\\ &R_\ell(\mathbf{r}_0, \mathbf{r}_0', \tau_3) dU_1\, dU_2\end{aligned} \tag{11.5}$$

where $R_1(\mathbf{r}_0, \mathbf{r}_0', \tau_3)$ is the cross-correlation function of the loading. Now, form the cross spectrum of the response:

$$S_w(\mathbf{r}_1, \mathbf{r}_2, \omega) = \int_{-\infty}^{+\infty} R_w(\mathbf{r}_1, \mathbf{r}_2, \tau_3) e^{-i\omega\tau} d\tau \tag{11.6}$$

$$e^{-i\omega\tau} = e^{-i\omega(\tau_3 - U_1 + U_2)}$$

Thus,

$$\begin{aligned} S_w(\mathbf{r}_1, \mathbf{r}_2, \omega) = &\int_{\mathbf{r}_0}\int_{\mathbf{r}_0'} d\mathbf{r}_0 d\mathbf{r}_0' \\ &\times \int_{-\infty}^{+\infty} h(\mathbf{r}_1, \mathbf{r}_0, U_1)\, e^{i\omega U_1} dU_1 \\ &\times \int_{-\infty}^{+\infty} h(\mathbf{r}_2, \mathbf{r}_0', U_2)\, e^{-i\omega U_2} dU_2 \\ &\times \left[\int_{-\infty}^{+\infty} R_\ell(\mathbf{r}_0, \mathbf{r}_0', \tau_3)\, e^{-i\omega\tau_3} d\tau_3\right] \end{aligned} \tag{11.7}$$

But the Fourier transform of the impulse function is the Green's function. Thus,

$$\begin{aligned} \int_{-\infty}^{+\infty} h(\mathbf{r}_1, \mathbf{r}_0, U_1) e^{i\omega U_1} dU_1 &= G^*(\mathbf{r}_1, \mathbf{r}_0, \omega) \\ \int_{-\infty}^{+\infty} h(\mathbf{r}_2, \mathbf{r}_0', U_2) e^{-i\omega U_2} dU_2 &= G(\mathbf{r}_2, \mathbf{r}_0', \omega) \end{aligned} \tag{11.8}$$

where

G^* denotes the complex conjugate of G

the Green's function $G(\mathbf{r}_2, \mathbf{r}_0, \omega)$ is the response at $\mathbf{r}_2$ due to a unit sinusoidal load at $\mathbf{r}_0$

$G^*(\mathbf{r}_1, \mathbf{r}_0, \omega)$ is the complex conjugate of the response at $\mathbf{r}_1$ due to a unit sinusoidal load at $\mathbf{r}_0$

The bracket can be written

$$S_\ell(\mathbf{r}_0, \mathbf{r}_0', \omega) = \int_{-\infty}^{+\infty} R_\ell(\mathbf{r}_0, \mathbf{r}_0', \tau_3) e^{-i\omega\tau_3} d\tau_3 \tag{11.9}$$

where S_1 is the cross spectrum of the load. Thus, the expression for the cross spectrum of the response becomes

$$S_w(\mathbf{r}_1, \mathbf{r}_2, \omega) = \int_{\mathbf{r}_0}\int_{\mathbf{r}_0'} d\mathbf{r}_0 d\mathbf{r}_0' G^*(\mathbf{r}_1, \mathbf{r}_0, \omega) G(\mathbf{r}_2, \mathbf{r}_0', \omega) S_\ell(\mathbf{r}_0, \mathbf{r}_0', \omega) \tag{11.10}$$

The spectrum at any point $\mathbf{r}_1$ is obtained by setting $\mathbf{r}_1 = \mathbf{r}_2$; thus,

$$S_w(\mathbf{r}_1, \omega) = \int_{\mathbf{r}_0}\int_{\mathbf{r}_0'} d\mathbf{r}_0 d\mathbf{r}_0' G^*(\mathbf{r}_1, \mathbf{r}_0, \omega) G(\mathbf{r}_1, \mathbf{r}_0', \omega) S_\ell(\mathbf{r}_0, \mathbf{r}_0', \omega) \tag{11.11}$$

Note the equivalence between Equation 11.10 and the general Equation 9.14 for linear systems.

In many practical cases, especially in turbulence excitation, the cross spectrum of the loading takes a homogeneous form, as follows:

$$S_\ell(\mathbf{r}_0, \mathbf{r}_0', \omega) = S(\mathbf{r}_0 - \mathbf{r}_0', \omega) \tag{11.12}$$

Let $\mathbf{r}_0 - \mathbf{r}_0' = \xi$. Equation 11.10 now becomes

$$S_w(\mathbf{r}_1, \mathbf{r}_2, \omega) = \int_{\mathbf{r}_0}\int_{\mathbf{r}_0'} d\mathbf{r}_0 d\mathbf{r}_0' G(\mathbf{r}_1, \mathbf{r}_0', \omega) G(\mathbf{r}_2, \mathbf{r}_0', \omega) S(\xi, \omega) \tag{11.13}$$

White[5] has shown that by applying Parseval's theorem and letting $\mathbf{r}_1 = \mathbf{r}_2$, the preceding equation can be written as

$$S_w(\mathbf{r}_1, \mathbf{r}_1', \omega) = (2\pi)^2 \int_{\mathbf{k}} \tilde{S}(\mathbf{k}, \omega) \psi(\mathbf{k}, \omega) d\mathbf{k} \tag{11.14}$$

where

$$\psi(\mathbf{k}, \omega) = \left| \frac{1}{(2\pi)^2} \int_{-\infty}^{+\infty} G(\mathbf{r}_1, \mathbf{r}, \omega) e^{ik(\mathbf{r}-\mathbf{r}_1)} d\mathbf{r} \right|^2 \tag{11.15}$$

In the preceding equations, $S(\mathbf{k}, \omega)$ is the spectrum of the excitation field in wave number space, and $\psi(\mathbf{k}, \omega)$ is the square of the Fourier transform of the Green's function, which can be obtained very quickly on a computer by application of the fast Fourier transform technique.

The Green's functions take on the true spatial character of an influence function. They represent the response at one point due to a unit sinuosidal load at another point. The inputs are loads, the outputs are deflections, and the linear black boxes are pieces of the structure as used in the first section of this chapter.

A few very interesting results can immediately be written from Equation 11.11. Supposing that a body is loaded by a single random force at point p, the loading $l(\mathbf{r}, t)$ can be written as

$$\ell(\mathbf{r}, t) = P(t)\delta(\mathbf{r} - \mathbf{r}_p) \tag{11.16}$$

The δ function signifies that 1 is 0 except when $\mathbf{r} = \mathbf{r}_\mathrm{p}$. Thus,

$$S_\ell(\mathbf{r}_0, \mathbf{r}_0', \omega) = S_p(\omega)\delta(\mathbf{r}_0 - \mathbf{r}_p)\delta(\mathbf{r}_0' - \mathbf{r}_p) \tag{11.17}$$

The spectrum of the response is therefore

$$\begin{aligned} &S_w(\mathbf{r}_1, \omega) \\ &\quad = \int_{\mathbf{r}_0}\int_{\mathbf{r}_0'} d\mathbf{r}_0 d\mathbf{r}_0' G^*(\mathbf{r}_1, \mathbf{r}_0, \omega) \\ &\quad G(\mathbf{r}_1, \mathbf{r}_0', \omega) S_p(\omega)\delta(\mathbf{r}_0 - \mathbf{r}_p)\delta(\mathbf{r}_0' - \mathbf{r}_p) \end{aligned} \tag{11.18}$$

$$\begin{aligned} &= S_p(\omega)\int_{\mathbf{r}_0} G^*(\mathbf{r}_1, \mathbf{r}_0, \omega) \\ &\quad \delta(\mathbf{r}_0 - \mathbf{r}_p) d\mathbf{r}_0 \int_{\mathbf{r}_0'} G(\mathbf{r}_1, \mathbf{r}_0', \omega) \\ &\quad \delta(\mathbf{r}_0' - \mathbf{r}_p) d\mathbf{r}_0' \end{aligned} \tag{11.19}$$

$$= S_p(\omega)\left|G(\mathbf{r}_1, \mathbf{r}_p, \omega)\right|^2 \tag{11.20}$$

The spectrum of the response is the square absolute value of the Green's function multiplied by the spectrum of the force. The Green's function in this case is the response at $\mathbf{r}_1$ due to unit sinusoidal force of frequency ω at $\mathbf{r}_\mathrm{p}$ (p being the loading point).

Suppose that there is a group of independent forces on the structure. The cross correlation between them is 0, so

$$S_\ell(\mathbf{r}_0,\ \mathbf{r}_0',\ \omega) = S(\mathbf{r}_0,\ \omega)\delta(\mathbf{r}_0 - \mathbf{r}_0') \tag{11.21}$$

That is, $S_1 = 0$ except when $\mathbf{r}_0 = \mathbf{r}_0'$, so

$$\begin{aligned} S_w(\mathbf{r}_1,\ \omega) &= \int_{\mathbf{r}_0}\int_{\mathbf{r}_0'} G^*(\mathbf{r}_1,\ \mathbf{r}_0,\ \omega)G(\mathbf{r}_1,\ \mathbf{r}_0',\ \omega) \\ &\quad S(\mathbf{r}_0,\ \omega)\delta(\mathbf{r}_0 - \mathbf{r}_0')\ d\mathbf{r}_0\ d\mathbf{r}_0' \\ &= \int_{\mathbf{r}_0'} G(\mathbf{r}_1,\ \mathbf{r}_0',\ \omega)\left[\int_{\mathbf{r}_0} G^*(\mathbf{r}_1,\ \mathbf{r}_0,\ \omega)\right. \\ &\quad \left. S(\mathbf{r}_0,\ \omega)\delta(\mathbf{r}_0 - \mathbf{r}_0')d\mathbf{r}_0\right]d\mathbf{r}_0' \\ &= \int_{\mathbf{r}_0'} G(\mathbf{r}_1,\ \mathbf{r}_0',\ \omega)\, G^*(\mathbf{r}_1,\ \mathbf{r}_0',\ \omega) \\ &\quad S(\mathbf{r}_0',\ \omega)]d\mathbf{r}_0' \\ &= \int_{\mathbf{r}_0'} \left|G(\mathbf{r}_1,\ \mathbf{r}_0',\ \omega)\right|^2 S(\mathbf{r}_0',\ \omega)d\mathbf{r}_0' \end{aligned} \tag{11.22}$$

If there are n forces, each with spectrum $S(\mathbf{r}_n,\ \omega)$,

$$S_w(\mathbf{r}_1,\ \omega) = \sum_n \left|G(\mathbf{r}_1,\ \mathbf{r}_n,\ \omega)\right|^2 S(\mathbf{r}_n,\ \omega)$$

The response is just the sum of the spectra for each force acting separately.

11.2 Computation of the Response in Terms of Modes

The general variational equation of motion for any elastic structure can be written as

$$\begin{aligned} &\iiint_V [\rho(\ddot{u}\ \delta u + \ddot{v}\ \delta v + \ddot{w}\ \delta w) + \delta W]dV \\ &- \iint_S (X_v \delta u + Y_v \delta v + Z_v \delta w)dS = 0 \end{aligned} \tag{11.23}$$

where
ρ is mass density of body
u, v, and w are displacements at any point
δu, δv, and δw are variations of the displacements
X_v, Y_v, and Z_v are surface forces
ds is the elemental surface area
dV is the elemental volume
δW is the variation of potential energy

In accordance with Love's analysis, let the displacements in the normal modes be described by

$$u = u_r\varphi_r', \qquad v = v_r\varphi_r', \qquad w = w_r\varphi_r' \tag{11.24}$$

where $\varphi_r' = A_r \cos p_r t$, p_r, which is the natural frequency of the rth mode. Now let the forced motion of the system be described by

$$u = \sum_r u_r\varphi_r, \qquad v = \sum_r v_r\varphi_r, \qquad w = \sum_r w_r\varphi_r \tag{11.25}$$

where u_r, v_r, and w_r are the mode shapes, and φ_r is a function of time. In accordance with Love, let

$$\begin{aligned} u &= u_r\varphi_r & \delta u &= u_s\varphi_s \\ v &= v_r\varphi_r & \delta v &= v_s\varphi_s \\ w &= w_r\varphi_r & \delta w &= w_s\varphi_s \end{aligned} \tag{11.26}$$

Substituting into the variational equation of motion, we obtain the following:

$$\begin{aligned} &\iiint_V \rho(u_r\ddot{\varphi}_r u_s\varphi_s + \upsilon_r\ddot{\varphi}_r\upsilon_s\varphi_s \\ &\qquad + w_r\ddot{\varphi}_r w_s\varphi_s)dV + \iiint_V \delta W\, dV \\ &= \iint_S (X_v u_s\varphi_s + Y_v\upsilon_s\varphi_s + Z_v w_s\varphi_s)\, dS \end{aligned} \tag{11.27}$$

However, the modal functions satisfy the equation for free vibration:

$$\iiint_V \delta W \, dV = \iiint_V \rho(p_r^2 u_r \varphi_r u_s \varphi_s + p_r^2 v_r \varphi_r v_s \varphi_s + p_r^2 w_r \varphi_r w_s \varphi_s) \, dV \tag{11.28}$$

and Love shows that

$$\iiint_V \rho(u_r u_s + v_r v_s + w_r w_s) \, dV = 0 \qquad r \neq s \tag{11.29}$$

Therefore, the final equation of motion becomes

$$\ddot{\varphi}_r(t) + p_r^2 \varphi(t) = F_r(t) \tag{11.30}$$

where

$$M_r = \iiint_V \rho(u_r^2 + v_r^2 + w_r^2) \, dV$$

the generalized mass for the rth mode.

$$F_r(t) = \frac{1}{M_r} \iint_S [X_v(t) u_r + Y_v(t) v_r + Z_v(t) w_r] \, dS \tag{11.31}$$

If structural damping is taken into account, it can be written as another generalized force that opposes the motion

$$(F_r)_{\text{damping}} = -\kappa \ddot{\varphi}_r \iiint_V (u_r^2 + v_r^2 + w_r^2) \, dV \tag{11.32}$$

where κ is the damping force per unit volume per unit velocity. Finally, the equation of motion becomes

$$\ddot{\varphi}_r + \psi_r \ddot{\varphi}_r + p_r^2 \varphi_r = F_r \tag{11.33}$$

where

$$\psi_r = \frac{\kappa}{M_r} \iiint_V (u_r^2 + v_r^2 + w_r^2) \, dV$$

It is convenient to employ the vector notation; thus, let the displacement functions in the rth mode be written as

$$\mathbf{q}_r = u_r\mathbf{i} + v_r\mathbf{j} + w_r\mathbf{k} \tag{11.34}$$

where **i**, **j**, and **k** are the unit vectors in the x-, y-, and z-directions, respectively. Let

$$\mathbf{F}(s,\ t) = X_v\mathbf{i} + Y_{v\mathrm{j}} + Z_v\mathbf{k} \tag{11.35}$$

Thus,

$$\begin{aligned} M_r &= \iiint\limits_V \rho\mathbf{q}_r \cdot \mathbf{q}_r dV, \qquad \mathbf{q}_r = \mathbf{q}_r(V) \\ F_r(t) &= \frac{1}{M_r}\iint\limits_S \mathbf{F}\cdot\mathbf{q}_r\, ds, \qquad \mathbf{F} = \mathbf{F}(S, t) \end{aligned} \tag{11.36}$$

The Fourier transform of $\mathbf{F}(S, t)$ is

$$S_{\mathbf{F}}(S,\ \omega) = \int_{-\infty}^{+\infty} \mathbf{F}(S,\ t)e^{-i\omega t}dt \tag{11.37}$$

and the Fourier transform of ϕ_r is

$$S_{\varphi_r}(\omega) = \int_{-\infty}^{+\infty} \varphi_r(t)e^{-i\omega t}dt \tag{11.38}$$

Now,

$$\begin{aligned} \varphi_r(t) &= \frac{1}{2\pi}\int_{-\infty}^{+\infty} S_{\varphi_r}(\omega)e^{i\omega t}d\omega \\ \dot{\varphi}_r(t) &= \frac{1}{2\pi}\int_{-\infty}^{+\infty} i\omega S_{\varphi_r}(\omega)e^{i\omega t}d\omega \\ \ddot{\varphi}_r(t) &= \frac{1}{2\pi}\int_{-\infty}^{+\infty} -\omega^2 S_{\varphi_r}(\omega)e^{i\omega t}d\omega \end{aligned} \tag{11.39}$$

$$S_{\dot{\varphi}_r}(\omega) = i\omega S_{\varphi_r}(\omega), \qquad S_{\ddot{\varphi}_r}(\omega) = -\omega^2 S_{\varphi_r}(\omega) \tag{11.40}$$

Let β_r be the damping constant for the rth mode. Now, take the Fourier transform of the equation of motion

$$S_{\ddot{\varphi}_r} + \beta_r S_{\dot{\varphi}_r} + p_r^2 S_{\varphi_r} = S_{F_r} \tag{11.41}$$

which is

$$-\omega^2 S_{\varphi_r}(\omega) + i\omega\beta_r S_{\varphi_r}(\omega) + p_r^2 S_{\varphi_r}(\omega) = S_{F_r}(\omega) \tag{11.42}$$

where

$$S_{F_r}(\omega) = \frac{1}{M_r} \iint_S S_{\mathbf{F}}(\mathbf{r}_s, \omega) \cdot \mathbf{q}_r(\mathbf{r}_s)\, ds \tag{11.43}$$

Therefore,

$$S_{\varphi_r}(\omega) = \frac{\iint_S S_{\mathbf{F}}(\mathbf{r}_s, \omega) \cdot \mathbf{q}_r(\mathbf{r}_s) ds}{M_r[(p_r^2 - \omega^2) + i\omega\beta_r]}$$

In dealing with statistical averaging, the cross-correlation function is used. The cross correlation between the displacement at two points in any direction (the direction can be different at the two points) is

$$\Gamma_q(\mathbf{r}_1, \mathbf{r}_2, \tau) = \lim_{T\to\infty} \frac{1}{2T} \int_{-T}^{+T} q(\mathbf{r}_1, t)\, q(\mathbf{r}_2, t+\tau)\, dt \tag{11.44}$$

We are picking a given direction at each point, so the two quantities are scalar (no longer vector). Then,

$$q = \sum_r q_r \varphi_r$$

$$q(\mathbf{r}_2, t+\tau) = \frac{1}{2\pi} \int_{-\infty}^{+\infty} S_q(\mathbf{r}_2, \omega)\, e^{i\omega(t+\tau)} d\omega \tag{11.45}$$

$$q(\mathbf{r}_1, t) = \frac{1}{2\pi} \int_{-\infty}^{+\infty} S_q(\mathbf{r}_1, \omega)\, e^{i\omega t} d\omega$$

Thus,

$$\Gamma_q(\mathbf{r}_1, \mathbf{r}_2, \tau) = \lim_{T\to\infty} \frac{1}{2\pi} \int_{-T}^{+T} q(\mathbf{r}_1, t) \times \left(\frac{1}{2\pi} \int_{-T}^{+T} S_q(\mathbf{r}_2, \omega) e^{i\omega(t+\tau)} d\omega \right) dt$$
$$= \lim_{T\to\infty} \frac{1}{2T} \int_{-T}^{+T} S_q(\mathbf{r}_2, \omega)\, e^{i\omega\tau} \times \left(\frac{1}{2\pi} \int_{-T}^{+T} q(\mathbf{r}_1, t) e^{i\omega t} dt \right) d\omega \tag{11.46}$$
$$= \lim_{T\to\infty} \frac{1}{2T} \int_{-T}^{+T} S_q^T(\mathbf{r}_2, \omega) \times S_q^{*T}(\mathbf{r}_1, \omega) e^{i\omega\tau} d\omega$$

Now, the power spectral density of the displacement is defined in terms of the cross correlation as

$$\Gamma_q(\mathbf{r}_1, \mathbf{r}_2, \tau) = \frac{1}{2\pi} \int_{-\infty}^{+\infty} G_q(\mathbf{r}_1, \mathbf{r}_2, \omega) e^{i\omega\tau} d\omega \tag{11.47}$$

Then,

$$G_q(\mathbf{r}_1, \mathbf{r}_2, \omega) = \lim_{T\to\infty} \frac{1}{2T} S_q^{*T}(\mathbf{r}_1, \omega) S_q^T(\mathbf{r}_2, \omega) \tag{11.48}$$

Now,

$$S_q^T(\mathbf{r}_2, \omega) = \sum_r q_r(\mathbf{r}_2) S_{\varphi_r}^T(\omega) \tag{11.49}$$

Thus,

$$G_q(\mathbf{r}_1, \mathbf{r}_2, \omega) = \lim_{T\to\infty} \frac{1}{2T} \sum_r \sum_k q_r(\mathbf{r}_1) S_{\varphi_r}^T(\omega) q_k(\mathbf{r}_2) S_{\varphi k}^T(\omega) \tag{11.50}$$

$$G_q(\mathbf{r}_1, \mathbf{r}_2, \omega) = \sum_r \sum_k \frac{q_r(\mathbf{r}_1) q_k(\mathbf{r}_2)}{Y_r^*(i\omega) Y_k(I\omega)} \iint_{S_u} \iint_{S_v}$$

$$\lim_{T \to \infty} \frac{1}{2T} [S_{\mathbf{F}}^{*T}(\mathbf{r}_{S_u}, \omega) \cdot \mathbf{q}_r(\mathbf{r}_{S_u})] \tag{11.51}$$

$$[S_F^T(\mathbf{r}_{S_v}, \omega) \cdot \mathbf{q}_k(\mathbf{r}_{S_v})] ds_u ds_v$$

Now if the integrand is written in the double surface integral, it is

$$\begin{aligned} &S_X^{*T} S_X^T u_r u_k + S_X^{*T} S_Y^T u_r v_k + S_X^{*T} S_Z^T u_r w_k \\ &+ S_Y^{*T} S_X^T v_r u_k + S_Y^{*T} S_Y^T v_r v_k + S_Y^{*T} S_Z^T v_r w_k \\ &+ S_Z^{*T} S_X w_r u_k + S_Z^{*T} S_Y^T w_r v_k + S_Z^{*T} S_Z^T w_r w_k \end{aligned} \tag{11.52}$$

Note the tensor of the properties of the last expression involving each component of loading. Note that, in the general formula involving the dot product, the component of the modal vector in the direction of the loading function at the two points has to be taken. Now, assuming that the loading is normal to the surface of the structure, our concern is with the cross-spectral density of the normal acceleration at two points (or cross-spectral density between normal acceleration at two points $\mathbf{r}_{1s}$ and $\mathbf{r}_{2s}$) on the surface:

$$a_n(\mathbf{r}_{1s}, \mathbf{r}_{2s}, \omega) = \omega^4 \sum_r \sum_q \frac{[q_r(\mathbf{r}_{1s}) q_k(\mathbf{r}_{2s})]_n}{Y_r^*(i\omega) Y_k(i\omega)}$$

$$\iint_{S_u} \iint_{S_v} G_p(\mathbf{r}_{S_u}, \mathbf{r}_{S_v}, \omega) \tag{11.53}$$

$$q_r(\mathbf{r}_{S_u}) q_k(\mathbf{r}_{S_v}) ds_u \, ds_v$$

where $G_p(\mathbf{r}_{S_u}, \mathbf{r}_{S_v}, \omega)$ is the cross-spectral density of the loading normal to the surface at points $\mathbf{r}_{S_u}$ and $\mathbf{r}_{S_v}$. The mean square acceleration over a frequency band Ω_1 to Ω_2 at point $\mathbf{r}_{1S}$ is given by

$$\overline{a_n(\mathbf{r}_{1s})_{\Delta\Omega}}^2 = \frac{1}{2\pi} \int_{\Omega_1}^{\Omega_2} a_n(\mathbf{r}_{1S}, \mathbf{r}_{1s}, \omega) \, d\omega \tag{11.54}$$

Equation 11.53 is nothing other than Equation 11.10 with the integrand expanded in terms of modes of the structure. In turn, Equation 11.10 is nothing other than Equation 9.14 written for a continuous structure instead of just a linear black box system.

11.3 Coupled Structural Acoustic Systems

Equation 10.8 stated that the cross-spectral density of the field pressure in terms of acceleration spectra on the surface is

$$G(P_1, P_2, \omega) = \frac{1}{(4\pi)^2} \int_{S_1} \int_{S_2} \rho_0^2 a_n(S_1, S_2, \omega) \tag{11.55}$$
$$\bar{g}(P_1, S_1, \omega)\bar{g}^*(P_2, S_2, \omega)\, dS_1\, dS_2$$

Furthermore, it was found in the last section that the cross-spectral density of the normal acceleration for a structure in which the loading is normal to the surface can then be written as

$$a_n(S_1, S_2, \omega) = \omega^4 \sum_r \sum_m \frac{q_{rn}(S_1)q_{mn}(S_2)}{Y_r(i\omega)Y_m^*(i\omega)} C_{rm}(\omega) \tag{11.56}$$

in which

$$C_{rm}(\omega) =$$
$$\int_{S_1} \int_{S_2} G(S_1, S_2, \omega)\, q_{rn}(S_1) q_{mn}(S_2)\, dS_1\, dS_2$$

where $G(S_1, S_2, \omega)$ is the cross-spectral density of the pressure that excites the structure, and $q_{rn}(S_1)$ is the normal component of the rth mode evaluated at point S_1 of the surface. If the damping in the structure is relatively low, then in accordance with the analysis of Powell[7] and Hurty and Rubenstein,[8] the cross-product terms can be neglected and

$$a_n(S_1, S_2, \omega) \approx \omega^4 \sum_r \frac{q_{rn}(S_1)q_{rn}(S_2)}{|Y_r(i\omega)|^2} C_{rr}(\omega) \tag{11.57}$$

where

$$C_{rr}(\omega) =$$
$$\int_{S_1} \int_{S_2} G(S_1, S_2, \omega) q_{rn}(S_1) q_{rn}(S_2)\, dS_1\, dS_2$$

To carry the analysis further, a Green's function must be obtained. Using the analysis of Strasberg[16] and Morse and Ingard[17] as a guide, we assume the use of a free-field Green's function. The analysis, although approximate, then comes out in general form instead of being limited to a particular surface. Therefore, let

$$\bar{g}(P_1, S_1, \omega) = \frac{e^{ikR_1}}{R_1} e^{-ik(\mathbf{a}R_1 \cdot \mathbf{R}S_1)} dS_2 \tag{11.58}$$

in which (see Figure 2.9)

$$\mathbf{a}_{R_1} \cdot \mathbf{R}_{S_1} = z_o \cos\theta_1 + x_o \sin\theta_1 \cos\varphi_1$$
$$+ y_o \sin\theta_1 \sin\varphi_1$$

where

x_o, y_o, and z_o are the rectangular coordinates of the point on the vibrating surface of the structure

$\mathbf{R}s_1$ is the radius vector to point S_1 on the surface

R_1, θ_1, and φ_1 are the spherical coordinates of point P_1 in the far field

$\mathbf{a}_{R_1}$ is a unit vector in the direction of R_1 (the radius vector from the origin to the far field point)

Thus, $\mathbf{a}_{R_1}$ $\mathbf{R}_{S_1}$ is the projection of $\mathbf{R}_{S_1}$ on R_1, making $\mathbf{R}_1 - \mathbf{a}_R \mathbf{R}_{S_1}$ the distance from the far field point to the surface point. Therefore,

$$\bar{g}^*(P_2, S_2, \omega) = \frac{e^{-ikR_2}}{R_2} e^{ik(\mathbf{a}R_2 \cdot \mathbf{R}_{S_2})} \tag{11.59}$$

Combining Equations 11.55, 11.56, 11.58, and 11.59 gives the following expression for far field cross spectrum of far field pressure:

$$G(P_1, P_2, \omega) = \omega^4 \rho_o^2 \frac{e^{ik}(R_1 - R_2)}{(4\pi)^2 R_1 R_2} \sum_r \sum_m$$
$$I_r(\theta_1, \varphi_1, \omega) I_m^*(\theta_2, \varphi_2, \omega) \times \frac{C_{rm}}{Y_r(i\omega) Y_m^*(i\omega)} \tag{11.60}$$

where

$$I_r = \int_{S_1} q_{rn}(S_1) e^{-ik(\mathbf{a}_{R_1} \cdot \mathbf{R}_{S_1})} dS_1$$

$$I_m^* = \int_{S_2} q_{mn}(S_2) e^{-ik(\mathbf{a}_{R_2} \cdot \mathbf{R}_{S_2})} dS_2 \tag{11.61}$$

With the low damping approximation given by Equation 11.57, the far field autospectrum at point P_i is

$$G(P_1, P_1, \omega) \approx \frac{\rho_o \omega^4}{(4\pi)^2 R_1^2} \times \sum_r \frac{\left|I_r(\theta_1, \varphi_1, \omega)\right|^2}{\left|Y_r(i\omega)\right|^2} C_{rr}(\omega) \tag{11.62}$$

The far field mean square pressure in a frequency band $\Delta\Omega = \Omega_2 - \Omega_1$ can be written as

$$\overline{p(P_1)^2_{\Delta\Omega}} = \frac{1}{2\pi} \int_{\Omega_1}^{\Omega_2} G(P_1, P_1, \omega)\, dw \tag{11.63}$$

Thus,

$$\overline{p(P_1)^2_{\Delta\Omega}} = \frac{\rho_o^2}{(4\pi)^2 R_1^2} \frac{1}{2\pi} \times \int_{\Omega_1}^{\Omega_2} \left[\sum_r \frac{\omega^4 C_{rr}(\omega)}{\left|Y_r(i\omega)\right|^2} \left|I_r(\theta_1, \varphi_1, \omega)\right|^2 dw \right. \tag{11.64}$$

In cases in which the structure is lightly damped, the following can be written:

$$\frac{1}{2\pi} \int_{\Omega_1}^{\Omega_2} \frac{\omega^4 C_{rr}(\omega) d\omega}{\left|Y_r(i\omega)\right|^2} \approx \frac{p_r C_{rr}(p_r)}{8\zeta_r M_r^2} \tag{11.65}$$

where $C_{rr}(p_r)$ is defined as the joint acceptance evaluated at the natural frequency p_r (p_r consists of those natural frequencies between Ω_1 and Ω_2), M_r is the total generalized mass of the rth mode (including virtual mass), and

$$\zeta_r = \bar{C}_r / (\bar{C}_c)_r \tag{11.66}$$

References

1. Morse, P. M., and Feshbach, H. 1953. *Methods of theoretical physics,* vol. 1, 1339. New York: McGraw–Hill Book Co., Inc.
2. Morse, P. M., and Feshbach, H. 1953. *Methods of theoretical physics,* vol. 1, 1343. New York: McGraw–Hill Book Co., Inc.
3. Yang, C. S., and Willmarth, W. W. 1969. Wall pressure fluctuations beneath an axially symmetric turbulent boundary layer on a cylinder, Dept. of Aerospace Engineering, University of Michigan, contract no. N00014-67-A-O181-0015.
4. Leibowitz, R. C. 1967. Turbulence-induced vibrations and radiations—A projection of the state of the art to naval research needs. Acoustics and Vibration Laboratory, NSRDC, tech. note AVL-185-942.
5. White, P. H. 1969. Application of the fast Fourier transform to linear distributed system response calculations. *Journal of Acoustical Society of America* 46:273.
6. Robson, J. D. 1964. *An introduction to random vibration.* New York: Elsevier Pub. Co.
7. Powell, A. 1958. On the response of structures to random pressures and to jet noise in particular. In *Random vibration,* ed. S. H. Crandall. New York: John Wiley & Sons, Inc.
8. Hurty, W. C., and Rubenstein, M. F. 1964. *Dynamics of structures,* chap. 11. Englewood Cliffs, NJ: Prentice Hall.
9. Love, A. E. H. 1944. *A treatise on the mathematical theory of elasticity,* 293. New York: Dover Publications.
10. Schroeder, M. R. 1969. Effect of frequency and space averaging on the transmission responses of multimode media. *Journal of Acoustical Society of America* 46 (2): 277.
11. Parrent, G. B., Jr. 1959. *Journal of Optical Society of America* 49:787.
12. Rayleigh, J. W. S. 1945. *The theory of sound.* New York: Dover Publications.
13. Marsh, H. W. 1951. Correlation in wave fields. USN Underwater Sound Lab. tech. mem. 921-54-51.
14. Powell, A. 1957. On structural vibration excited by random pressures with reference to structural fatigue and boundary layer noise. Douglas Aircraft Co., report no. SM-22795.
15. Schenk, H. A. 1968. Improved integral formulation for acoustic radiation problems. *Journal of Acoustical Society of America* 44 (1): 41.
16. Strasberg, M. 1962. Sound radiation from slender bodies in axisymmetric vibration. *Fourth International Congress on Acoustics,* Copenhagen, Aug. 21–28, 1962, paper 0-28.
17. Morse, P. M., and Ingard, K. U. 1961. Linear acoustic theory. In *Encyclopedia of physics,* vol. XI/1, Acoustics I. New York: Springer–Verlag.
18. Greenspon, J. E. 1992. Acoustics, linear. In *Encyclopedia of physical science and technology,* vol. 1, 114–152. New York: Academic Press, Inc.–Elsevier.

12

Random Radiation from Cylindrical Structures

12.1 Directivity Patterns

Many practical structures are composed of bodies of revolution containing bulkheads. These structures have complicated mass distribution due to machinery and internal equipment. However, each normal component of a natural mode of the structure can be approximated by ($r \to p, q$):

$$q_{rn}(s) \equiv q_{rn}(z,\varphi) = (A_{pq}\cos q\varphi + B_{pq}\sin q\varphi)\sin\frac{p\pi z}{\ell} \tag{12.1}$$

Expression 12.1 is a half-range Fourier expansion in the longitudinal direction (which is chosen as the distance along the surface of the structure) between bulkheads, which are distance ℓ apart, and a full-range Fourier expansion in the peripheral (φ) direction. The boundary conditions at the bulkheads will not be satisfied exactly by the sin ($p\pi z/\ell$) term, but Arnold and Warburton[1] have demonstrated that if an equivalent length is used, sin ($p\pi z/\ell$) can be used to approximate the mode shapes.

A great deal of the random vibration in hull type structures comes from these lateral motions of the hull or the cylindrical plating between ring stiffeners. Thus, for a cylindrical section of length ℓ, we have

$$I_r(\theta,\varphi_1,\omega) = -\int_0^{\ell}\int_0^{2\pi}\left\{(A_{pq}\cos q\varphi + B_{pq}\sin q\varphi)\sin\frac{p\pi z}{\ell}\right\} \left\{e^{-ikz\cos\theta_1}e^{-ika\sin\theta_1(\cos\varphi\cos\varphi_1+\sin\varphi\sin\varphi_1)}\right\}ad\varphi\ell z \tag{12.2}$$

Thus,

$$I_r = A_{pq}\left\{\int_0^{\ell}\cos(kz\cos\theta_1)\sin\frac{p\pi z}{\ell}dz - i\int_0^{\ell}\sin(kz\cos\theta_1)\sin\frac{p\pi z}{\ell}dz\right\}$$

$$\left\{\int_0^{2\pi}\cos q\varphi e^{-ika\sin\theta,\cos(\varphi-\varphi_1)}ad\varphi\right\}$$

$$+B_{pq}\left\{\int_0^{\ell}\cos(kz\cos\theta_1)\sin\frac{p\pi z}{\ell}dz - i\int_0^{\ell}\sin(kz\cos\theta_1)\sin\frac{p\pi z}{\ell}dz\right\}$$

$$\left\{\int_0^{2\pi}\sin q\varphi \varepsilon^{-ika\sin\theta,\cos(\varphi-\varphi_1)}ad\varphi\right\}$$

The integrals on z can be evaluated analytically without difficulty; however, the integrals on φ will involve use of the following result given by Smith and Kerwin[2]:

$$\int_0^{2\pi}\cos n\varphi e^{iz\cos(\varphi-\alpha)}d\varphi = 2\pi i^n \cos n\alpha J_n(z) \tag{12.3}$$

Likewise, using a proof similar to that of Smith and Kerwin,

$$\int_0^{2\pi}\sin n\varphi e^{iz\cos(\varphi-\alpha)}d\varphi = 2\pi i^n \sin n\alpha J_n(z)$$

Carrying out the integration and substituting into Equation 12.2, we finally obtain

$$\begin{aligned}
I_r(\theta_1,\varphi_1,\omega) = A_{pq}&\left\{\left[\frac{1}{2(p\pi-k\ell\cos\theta_1)}+\frac{1}{2(p\pi+k\ell\cos\theta_1)}\right]\right.\\
&-\left[\frac{\cos(p\pi-k\ell\cos\theta_1)}{2(p\pi-k\ell\cos\theta_1)}+\frac{\cos(p\pi+k\ell\cos\theta_1)}{2(p\pi+k\ell\cos\theta_1)}\right]\\
&\left.-i\left[\frac{\sin(p\pi-k\ell\cos\theta_1)}{2(p\pi-k\ell\cos\theta_1)}-\frac{\sin(p\pi+k\ell\cos\theta_1)}{2(p\pi+k\ell\cos\theta_1)}\right]\right\}2\pi a\ell i^q\cos q\varphi, J_q(-k_a\sin\theta_1)\\
+B_{pq}&\left\{\left[\frac{1}{2(p\pi-k\ell\cos\theta_1)}+\frac{1}{2(p\pi+k\ell\cos\theta_1)}\right]\right.\\
&-\left[\frac{\cos(p\pi-k\ell\cos\theta_1)}{2(p\pi-k\ell\cos\theta_1)}+\frac{\cos(p\pi+k\ell\cos\theta_1)}{2(p\pi+k\ell\cos\theta_1)}\right]\\
&\left.-i\left[\frac{\sin(p\pi-k\ell\cos\theta_1)}{2(p\pi-k\ell\cos\theta_1)}-\frac{\sin(p\pi+k\ell\cos\theta_1)}{2(p\pi+k\ell\cos\theta_1)}\right]\right\}2\pi a\ell i^q\sin q\varphi_1 J_q(-ka\sin\theta_1)
\end{aligned} \tag{12.4}$$

Consider the directivity pattern in the horizontal plane of the cylindrical structure—that is, at $\varphi_1 = \pi/2$ (see Figure 2.9 in Chapter 2) and then examine the pq^{th} term in the series of Equation 12.4. After some algebraic manipulation of Equation 12.4, we find the amplitude of $I_{pq}/2\pi a\ell$, which we denote by D, to be

$$\text{For } p\pi \neq k\ell\cos\theta_1 \quad D = \left|\frac{I_{pq}(\theta_1, \pi/2, \omega)}{2\pi a\ell}\right| = \left|\frac{\sqrt{2}p\pi[J_q(ka\sin\theta_1)\sqrt{1-(-1)^p\cos(k\ell\cos\theta_1)}}{(p\pi)^2 - (k\ell\cos\theta_1)^2}\right| \tag{12.5}$$

$$\text{For } p\pi = k\ell\cos\theta_1 \qquad D = \sqrt{\frac{p\pi}{2}}\; J_q(ka\sin\theta_1) \tag{12.6}$$

This directivity factor is plotted in Figure 2.10 (Chapter 2) for some representative cases.

Within the assumption that a free space Green's function can be used, Equation 11.60 is the general expression for the cross-spectral density of the far field sound pressure radiated from a randomly vibrating structure of any shape. The radial dependence is of the form $e^{ik(R_1-R_2)}/R_1R_2$ and therefore satisfies the Sommerfield radiation condition. The directivity is contained in the functions $I_r\,(\theta_1, \ell_1, \omega)$ and $I^*_m(\theta_2, \ell_2, \omega)$. The random loading and its coupling to the structure are contained in the joint acceptance function $C_{rm}(\omega)$ and the structural characteristics, such as damping and natural frequency, are contained in the frequency response functions $Y_r(i\omega)$ and $Y^*_m(i\omega)$.

Calculations based upon Equation 11.60 are certainly well within the capacity of today's computers. However, in order to gain further insight into the problem, cases in which product terms can be neglected have been determined. This simplifies the problem many times over, as can be seen from Equation 11.62. If, then, the special case of the cylindrical structure is considered, it is seen that a closed solution can be obtained for I_r. For frequency bands in which the natural frequencies are well separated, Equation 11.65 gives the mean square far field sound pressure.

For cylindrical structures, Figure 2.10 (in Chapter 2) gives the directivity patterns in the horizontal plane $\varphi_1 = \pi/2$ for various vibration shapes of the form $\cos q\varphi \sin(p\pi z/\ell)$. At very small ka (i.e., $ka = 0.1$), it is seen that for shapes in which $q = 0$, odd p terms are source type and even p terms give dipole type radiation. For $q = 1$, the odd p terms give dipole type radiation and the even p terms give the quadrupole type (see Figure 2.10). Note that at $ka = 0.1$, the magnitude of the largest dipole term is only a small fraction of the largest source term; likewise, the magnitude of the largest quadrapole type term is only a small fraction of the largest dipole contribution. At these low frequencies, the $q = 0$ terms are so much larger than terms with $q = 1, 2$ that we can characterize a cylindrical structure vibrating at frequencies in which $ka < 0.1$ as a simple source. (This, of course, assumes that the structural mode contributions from the $q = 0$, $p = 1, 3$, etc. terms are not much smaller than the other terms $q = 1$, $p = 1, 3$, etc.)

As the frequency is increased to $ka = 1.0$, it is seen that the dipole and quadrapole type terms become much more significant. As the frequency is increased further to $ka = 3$, the appearance of minor lobes becomes apparent; as the frequency and q are increased, the sound becomes highly directive to the broadside of the cylinder. For example, if we examine the curves for $ka = 3$, $q = 5$, it is seen that all of the sound is essentially radiated within a 60° band off the broadside of the cylinder.

12.2 Multipole Expansion

Using the low-frequency multipole expansion as given in Morse and Ingard,[9]

$$\int_{s_1} \bar{q}_{rn}(s_1)e^{-ik(\vec{a}_{R_1}\cdot\vec{R}_{s_1})}ds_1 = \int_{s_1} \bar{q}_{rn}(s_1)ds_1 - ik\vec{a}_{R_1}\bullet\left(\int_{s_1} \vec{R}_{s_1}\bar{q}_{rn}(s_1)ds_1\right) - \frac{1}{2}k^2\vec{a}_{R_1}\bullet\left(\int_{s_1} \vec{R}_{s_1}\bar{q}_{rn}(s_1)\vec{R}_{s_1}ds_1\right)\bullet\vec{a}_{R_1} \tag{12.7}$$

and a similar expression for the integral over S_2, these expressions can be written in terms of multipole moments as follows:

$\int_{S_1} \bar{q}_{rn}(s_1)ds_1 = V_r$ the volume change of the rth mode (monopole)

$\int_{S_1} \vec{R}_{S_1},\bar{q}_{rn}(s_1)ds_1 = D_r$ the dipole moment of the rth mode

$\int_{S_1} \vec{R}_{s_1}\bar{q}_{rn}(s_1)\vec{R}_{s_1}ds_1 = Q_r$ the quadrapole moment of the rth mode

Thus,

$$G(P_1,P_2,\omega)=\frac{\varepsilon^{ik}(R_1-R_2)}{(4\pi)^2R_1R_2}\rho_o^2\omega^4\sum_r\sum_m\frac{C_{rm}(\omega)}{Y_r(i\omega)Y_m^*(i\omega)}[\bar{V}_r-ik\bar{D}_r-\tfrac{1}{2}k^2\bar{Q}_r+\cdots\cdot]$$
$$[\bar{V}_m+ik\bar{D}_m-\tfrac{1}{2}k^2\bar{Q}_m+\cdots\cdot] \tag{12.8}$$

where the bars over the *V*, *D*, and *Q* denote that these quantities are the complete contributions from the respective multipoles after the vectorial operations have been performed on the multipole moment. Each normal component of a natural mode of the structure can be approximated, as shown before.

$$q_{rn}(S)=q_{rn}(\xi,\varphi)\approx(A_{pq}\cos q\varphi+B_{pq}\sin q\varphi)\sin\frac{p\pi\xi}{\ell} \tag{12.9}$$

For this set of mode shapes,

$$\bar{V}_r=\int_0^{2\pi}\int_0^{\ell}(A_{pq}\cos q\varphi+B_{pq}\sin q\varphi)\sin\frac{p\pi\xi}{\ell}r(\xi)d\varphi d\xi$$

where $r(\xi)$ is the radius of the body of revolution at point ξ. Carrying out the integration, it is seen that the only contribution to $\bar{V}_r$ is for $n = 0$—that is,

$$\bar{V}_r\approx A_{po}\int_0^{\ell}r(\xi)\sin(p\pi\xi/\ell)d\xi \tag{12.10}$$

The dipole contribution becomes

$$\bar{D}_r\approx\int_0^{\ell}\int_0^{2\pi}(A_{pq}\cos q\varphi+B_{pq}\sin q\varphi)\sin\frac{p\pi\xi}{\ell}$$
$$\times[z(\xi)\cos\theta_1+r(\xi)\cos\varphi\sin\theta_1\cos\varphi_1+r(\xi)\sin\varphi\sin\theta_1\sin\varphi_1]r(\xi)d\varphi d\xi \tag{12.11}$$

r, φ, and z are cylindrical coordinates; note that r and z are both functions of ξ, the distance along the surface (θ, φ refer to the far field point).

Carrying out the φ integration,

$$\begin{aligned}\bar{D}_r \approx \cos\theta_1 A_{po}\int_0^{\ell} z(\xi)r(\xi)\sin\frac{p\pi\xi}{\ell}d\xi \\ +\sin\theta_1\cos\varphi_1\sum_p A_{p1}\pi\int_0^{\ell} r^2(\xi)\sin\frac{p\pi\xi}{\ell}d\xi \\ +\sin\theta_1\sin\varphi_1\sum_p B_{p1}\pi\int_0^{\ell} r^2(\xi)\sin\frac{p\pi\xi}{\ell}d\xi\end{aligned} \tag{12.12}$$

and the quadrupole contribution becomes

$$\begin{aligned}\bar{Q}_r \approx \int_0^{2\pi}\int_0^{\ell}\left[(A_{pq}\cos q\varphi + B_{pq}\sin q\varphi\sin\frac{p\pi\xi}{\ell}\right] \\ \left[z^2(\xi)\cos^2\theta_1 + 2z(\xi)r(\xi)\cos\theta_1\sin\theta_1\cos\varphi_1\cos\varphi\right. \\ +2z(\xi)r(\xi)\cos\theta_1\sin\theta_1\sin\varphi_1\sin\varphi + r^2(\xi)\sin^2\theta_1\cos^2\theta_1\left(\frac{1+\cos 2\varphi}{2}\right) \\ \left.+2r^2(\xi)\sin^2\theta_1\cos\varphi_1\sin\varphi_1\frac{\sin 2\varphi}{2} + r^2(\xi)\sin^2\theta_1\sin^2\varphi_1\left(\frac{1-\cos 2\varphi}{2}\right)\right] r(\xi)d\varphi d\xi\end{aligned} \tag{12.13}$$

From the preceding expressions, it is plainly seen that the $q = 0$ structural components are the only ones that contribute to the monopole. The $q = 0, 1$ components contribute to the dipole and the $q = 0, 1, 2$ components contribute to the quadrupole, etc.

12.3 Random Loading and Response

12.3.1 Far Field from Random Loading

The far field pressure for a cylindrical shell under random loading can be written as[3]

$$p_{ff}(R,\varphi,\psi,t) = \sum_{n=0}^{\infty}\sum_{m=1}^{\infty} v_{mn}(R,\psi,t)\cos n\varphi + u_{mn}(R,\psi,t)\sin n\varphi \tag{12.14}$$

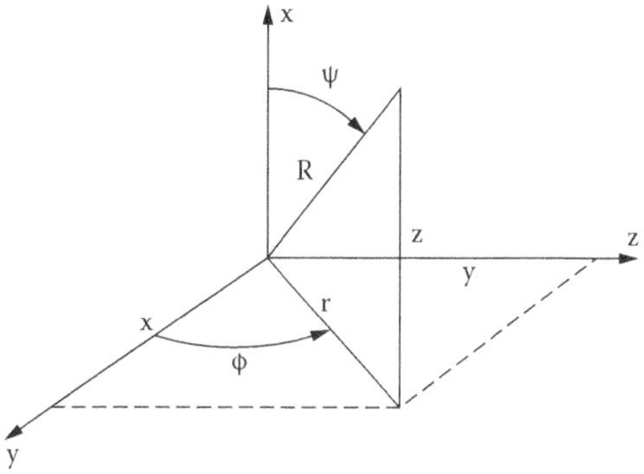

FIGURE 12.1
Coordinate system.

where R, θ, and ψ are spherical coordinates as shown in Figure 12.1.

In this representation of the far field pressure, it has been assumed that the loading on the shell has been decomposed into Fourier components as follows:

$$p(x,\varphi,t) = \sum_{n=0}^{\infty}\sum_{m=1}^{\infty}[f_{mn}(t)\cos n\varphi + g_{mn}(t)\sin n\varphi]\sin\frac{m\pi x}{\ell} \tag{12.15}$$

where

$$\begin{aligned} f_{mn}(t) &= \frac{2}{\pi\ell}\int_0^{\ell}\int_0^{2\pi} p(x,\varphi,t)\cos n\varphi \sin\frac{m\pi x}{\ell} dx d\varphi \\ g_{mn}(t) &= \frac{2}{\pi\ell}\int_0^{\ell}\int_0^{2\pi} p(x,\varphi,t)\sin n\varphi \sin\frac{m\pi x}{\ell} dx d\varphi \end{aligned} \tag{12.16}$$

The solution (Equation 12.14) for the far field pressure is based upon the fact that the displacement w of the cylindrical shell under random loading can be written as[3] (this is a particular solution of a cylindrical shell with freely supported ends):

$$w(x,q,t) = \sum_{n=0}^{\infty}\sum_{m=1}^{\infty}[\bar{v}_{mn}(t)\cos n\varphi + \bar{u}_{mn}(t)\sin n\varphi]\sin\frac{m\pi x}{\ell} \tag{12.17}$$

Thus, $v_{mn}(R,\psi,t)\cos n\varphi$ is the far field pressure arising for a shell displacement of $\bar{v}_{mn}\cos n\varphi \sin\frac{m\pi x}{\ell}$.

The far field functions v_{mn} and u_{mn} can be written in terms of impulse functions $h_{p_{ff}}$ and $k_{p_{ff}}$ as follows:

$$v_{mn}(R,\psi,t)=\int_{-\infty}^{+\infty} f_{mn}(u)h_{p_{ff}}(R,m,n,\psi,t-u)du$$
$$u_{mn}(R,\psi,t)=\int_{-\infty}^{+\infty} g_{mn}(u)k_{p_{ff}}(R,m,n,\psi,t-u)du \tag{12.18}$$

If the loading component has the form

$$p_h(x,\theta,t)=\cos n\varphi \sin\frac{m\pi x}{\ell}e^{i\omega t} \tag{12.19}$$

then the far field pressure will have the form

$$(p_{ff})_h = H_{p_{ff}}(R,m,n,\psi,\omega)\cos n\varphi e^{i\omega t} \tag{12.21}$$

and, if the loading has the form of

$$p_{\kappa}(x,\theta,t)=\sin n\varphi \sin\frac{m\pi x}{\ell}e^{i\omega t} \tag{12.22}$$

then the far field pressure has the form

$$(p_{ff})_k = K_{p_{ff}}(R,m,n,\psi,\omega)\sin\varphi e^{i\omega t} \tag{12.23}$$

where H and K are frequency response functions and are related to the impulse function by the Fourier transform—that is,

$$H_{p_{ff}}(R,m,n,\psi,\omega)=\int_{-\infty}^{+\infty} h_{p_{ff}}(R,m,n,\psi,t)e^{i\omega t}dt$$
$$K_{p_{ff}}(R,m,n,\psi,\omega)=\int_{-\infty}^{+\infty} k_{p_{ff}}(R,m,n,\psi,t)e^{i\omega t}dt \tag{12.24}$$

The cross spectrum of the far field pressure can then be written as follows:

$$\tilde{p}_{ff}(R_1,R_2,\varphi_1,\varphi_2,\psi_1,\psi_2,\omega)=\int_{-\infty}^{+\infty}\langle p_{ff}(R_1,\varphi_1,\psi_1,t)p_{ff}(R_2,\varphi_2,\psi_2,t+\tau)\rangle e^{-i\omega\tau}d\tau$$

$$\langle A\rangle=\lim_{T\to\infty}\frac{1}{2T}\int_{-T}^{+T}Adt \tag{12.25}$$

Substituting Equation 12.14,

$$\begin{aligned}
&\tilde{p}_{ff}(R_1,R_2,\varphi_1,\varphi_2,\psi_1,\psi_2,\omega)\\
&=\sum_{m=1}^{\infty}\sum_{p=1}^{\infty}\sum_{q=0}^{\infty}\sum_{n=0}^{\infty}\int_{-\infty}^{+\infty}\big[\langle v_{mn}(R_1,\psi_1,t)v_{pq}(R_2,\psi_2,t+\tau)\rangle\cos n\varphi_1\cos q\varphi_2\\
&+\langle u_{mn}(R_1,\psi_1,t)u_{pq}(R_2,\psi_2,t+\tau)\rangle\sin n\varphi_1\sin q\varphi_2\\
&+\langle u_{mn}(R_1,\psi_1,t)v_{pq}(R_2,\psi_2,t+\tau)\rangle\sin n\varphi_2\cos q\varphi_2\\
&+\langle v_{mn}(R_1,\psi_1,t)u_{pq}(R_2,\psi_2,t+\tau)\rangle\cos n\varphi_1\sin q\varphi_2\big]\\
&\times e^{-i\omega\tau}d\tau
\end{aligned} \tag{12.26}$$

where

$$\begin{aligned}
I_{mn}&=\int_{-\infty}^{+\infty}\langle v_{mn}(R_1,\psi_1,t)v_{pq}(R_2,\psi_2,t+\tau)\rangle\, e^{-i\omega\tau}d\tau\\
&=\int_{-\infty}^{+\infty}\left\{\frac{1}{2\tau}\int_{-\tau}^{+\tau}\left[\int_{-\infty}^{+\infty}f_{mn}(t_1)hp_{ff}(m,n,R_1,\psi_1,t-t_1)dt_1\right]\right.
\end{aligned} \tag{12.27}$$

where

$$\left.\left[\int_{-\infty}^{+\infty}f_{pq}(t_2)hp_{ff}(p,q,R_2,\psi_2,t-t_2+\tau)dt_2\right]dt\right\}e^{-i\omega\tau}d\tau$$

$$f_{mn}(t_1)=\frac{2}{\pi\ell}\int_0^{\ell}\int_0^{2\pi}p(x,\varphi,t)\cos n\varphi\sin\frac{m\pi x}{\ell}dxd\varphi \tag{12.28}$$

The assumption is now made that the process is stationary, so the loading will be a function of the difference of the times t_1 and t_2; let this difference be denoted by τ_3 and thus let

$$u_1 = t - t_1 \qquad u_2 = t - t_2 + \tau$$
$$\tau_3 = t_2 - t_1 = (t - u_2 + \tau) - (t - u_1) = u_1 - u_2 + \tau \tag{12.29}$$

Therefore,

$$I_{mn} = \frac{4}{\pi^2 \ell^2} \int_0^{\ell} \sin \frac{m\pi x_1}{\ell} dx_1 \int_0^{\ell} \sin \frac{p\pi x_2}{\ell} dx_2 \int_0^{2\pi} \cos n\varphi_1 d\varphi_1 \int_0^{2\pi} \cos q\varphi_2 d\varphi_2$$
$$\left\{ \int_{-\infty}^{+\infty} d\tau_3 \int_{-\infty}^{+\infty} du_1 \int_{-\infty}^{+\infty} du_2 h_{p_{ff}}(m,n,R_1,\psi_1,u_1) h_{p_{ff}}(p,q,R_2,\psi_2,u_2) e^{-i\omega(\tau_3 - u_1 + u_2)} \right.$$
$$\left. \left[\lim_{T\to\infty} \frac{1}{2T} \int_{-T}^{+T} p(x_1,\varphi_1,t-u_1) p(x_2,\varphi_2,t-u_2+\tau) dt \right] \right\} \tag{12.30}$$

Finally,

$$I_{mn} = \frac{4}{\pi^2 \ell^2} \int_0^{\ell} \sin \frac{m\pi x_1}{\ell} dx_1 \int_0^{\ell} \sin \frac{p\pi x_2}{\ell} dx_2 \int_0^{2\pi} \cos n\varphi_1 d\varphi_2 \int_0^{2\pi} \cos q\varphi_2 d\varphi_2$$
$$\left[\int_{-\infty}^{+\infty} h_{p_{ff}}(m,n,R_1,\psi_1,u_1) e^{i\omega u_1} du_1 \right] \left[\int_{-\infty}^{+\infty} h_{p_{ff}}(p,q,R_2,\psi_2,u_2) e^{-i\omega u_2} du_2 \right] \tag{12.31}$$
$$\left\{ \int_{-\infty}^{+\infty} \left[\lim_{T\to\infty} \frac{1}{2T} \int_{-T}^{+T} p(x_1,\varphi_1,t-u_1) p(x_2,\varphi_2,t-u_2+\tau) dt \right] e^{-i\omega\tau_3} d\tau_3 \right\}$$

Employing the equations for the frequency response functions as given by Equation 12.24,

$$I_{mn} = \frac{4}{\pi^2 \ell^2} \int_0^{\ell} \sin \frac{m\pi x_1}{\ell} dx_1 \int_0^{\ell} \sin \frac{p\pi x_2}{\ell} dx_2 \int_0^{2\pi} \cos n\varphi_1 d\varphi_1 \int_0^{2\pi} \cos q\varphi_2 d\varphi_2$$
$$\left[H_{p_{ff}}(m,n,R_1,\psi_1,\omega) \right] \left[H^*_{p_{ff}}(p,q,R_2,\psi_2,\omega) \right] \left[S_f(x_1,\varphi_1,x_2,\varphi_2,\omega) \right] \tag{12.32}$$

$$S_f(x_1,\varphi_1,x_2,\varphi_2,\omega)=\int_{-\infty}^{+\infty}\left[\lim_{T\to\infty}\frac{1}{2T}\int_{-T}^{+T}p(x_1,\varphi_1,t-u_1)p(x_2,\varphi_2,t-u_2+\tau)dt\right]e^{-i\omega\tau_3}d\tau_3$$

$$\begin{aligned}I_{mn}&=\int_{-\infty}^{+\infty}\langle v_{mn}(R_1,\psi_1,t)v_{pq}(R_2,\psi_2,t+\tau)\rangle e^{-i\omega\tau}d\tau\\&=H_{p_{ff}}(m,n,R_1,\psi_1,\omega)H^*(p,q,R_2,\psi_2,\omega)C_{v_{mn}v_{pq}}\\C_{v_{mn}v_{pq}}&=\frac{4}{\pi^2\ell^2}\int_0^{\ell}\int_0^{\ell}\int_0^{2\pi}\int_0^{2\pi}S_f(\xi_1,\eta_1,\xi_2,\eta_2,\omega)\sin\frac{m\pi\xi_1}{\ell}\\&\quad\sin\frac{p\pi\xi_2}{\ell}\cos n\eta\cos q\eta_2 d\xi_1 d\eta_1 d\xi_2 d\eta_2\end{aligned}\tag{12.33}$$

$$\begin{aligned}I_{mn}&=\int_{-\infty}^{+\infty}\langle u_{mn}(R_1,\psi_1,t)u_{pq}(R_2,\psi_2,t+\tau)\rangle e^{-i\omega\tau}d\tau\\&=H_{p_{ff}}(m,n,R_1,\psi_1,\omega)K^*_{p_{ff}}(p,q,R_2,\psi_2,\omega)C_{u_{mn}u_{pq}}\\C_{u_{mn}u_{pq}}&=\frac{4}{\pi^2\ell^2}\int_0^{\ell}\int_0^{\ell}\int_0^{2\pi}\int_0^{2\pi}S_f(\xi_1,\eta_1,\xi_2,\eta_2,\omega)\sin\frac{m\pi\xi_1}{\ell}\\&\quad\sin\frac{p\pi\xi_2}{\ell}\cos n\eta_1\cos q\eta_2 d\xi_1 d\eta_1 d\xi_2 d\eta_2\end{aligned}\tag{12.34}$$

$$\begin{aligned}&\int_{-\infty}^{+\infty}\langle u_{mn}(R_1,\psi_1,t)v_{pq}(R_2,\psi_2,t+\tau)\rangle e^{-i\omega\tau}d\tau\\&\quad=K_{p_{ff}}(m,n,R_1,\psi_1,\omega)H^*_{p_{ff}}(p,q,R_2,\psi_2,\omega)C_{u_{mn}v_{pq}}\\C_{u_{mn}v_{pq}}&=\frac{4}{\pi^2\ell^2}\int_0^{\ell}\int_0^{\ell}\int_0^{2\pi}\int_0^{2\pi}S_f(\xi_1,\eta_1,\xi_2,\eta_2,\omega)\sin\frac{m\pi\xi_1}{\ell}\\&\quad\sin\frac{p\pi\xi_2}{\ell}\sin n\eta_1\cos q\eta_2 d\xi_1 d\eta_1 d\xi_2 d\eta_2\end{aligned}\tag{12.35}$$

$$\int_{-\infty}^{+\infty}\langle v_{mn}(R_1,\psi_1,t)u_{pq}(R_2,\psi_2,t+\tau)\rangle e^{-i\omega\tau}d\tau$$

$$= H_{p_{ff}}(m,n,R_1,\psi_1,\omega)K^*_{p_{ff}}(p_1,q_1,R_2,\psi_2,\omega)C_{v_{mn}u_{pq}}$$

$$C_{v_{mn}u_{pq}} = \frac{4}{\pi^2\ell^2}\int_0^{\ell}\int_0^{\ell}\int_0^{2\pi}\int_0^{2\pi} S_f(\xi_1,\eta_1,\xi_2,\eta_2,\omega)\sin\frac{m\pi\xi_1}{\ell}$$
$$\sin\frac{p\pi\xi_2}{\ell}\sin q\eta_1\cos n\eta_2 d\xi_1 d\eta_1 d\xi_2 d\eta_2 \tag{12.36}$$

However,

$$K_{p_{ff}}(m,n,R_1,\psi_1,\omega) = H_{p_{ff}}(m,n,R_1,\psi_1,\omega) \tag{12.37}$$

H and K are the portions of the far field pressures dependent on R and ψ and independent of φ.

Now take the average far field pressure spectrum over φ, letting $R_1 = R_2 = R$; $\psi_1 = \psi_2 = \psi$; $\varphi_1 = \varphi_2 = \varphi$:

$$[\tilde{p}_{ff}(R,\psi,\omega)]_{Ave.\varphi} = \frac{1}{2\pi a}\sum_{m=1}^{\infty}\sum_{p=1}^{\infty}\sum_{n=0}^{\infty}\int_{-\infty}^{+\infty}[\langle v_{mn}(R,\psi,t)v_{pn}(R,\psi,t+\tau)\rangle\pi a$$
$$+\langle u_{mn}(R,\psi,t)u_{pn}(R,\psi,t+\tau)\rangle\pi a] \tag{12.38}$$
$$\times e^{-i\omega\tau}d\tau$$

Neglecting cross-product terms in m and p (neglecting cross-product terms is strictly legitimate if the natural frequencies are well separated and if the damping is low), we obtain

$$[\tilde{p}_{ff}(R,\psi,\omega)]_{Ave.\theta} = \frac{1}{2}\sum_m\sum_n\int_{-\infty}^{+\infty}[\langle v_{mn}(R,\psi,t)v_{mn}(R,\psi,t+\tau)\rangle \tag{12.39}$$
$$+\langle u_{mn}(R,\psi,t)u_{mn}(R,\psi,t+\tau)\rangle]e^{-i\omega\tau}d\tau$$

where

$$\int_{-\infty}^{+\infty}\langle v_{mn}(R,\psi,t)v_{mn}(R,\psi,t+\tau)\rangle e^{-i\omega\tau}d\tau = \left|H_{p_{ff}}(m,n,R,\psi,\omega)\right|^2 C_{v_{mn}v_{mn}} \tag{12.40}$$

$$\int_{-\infty}^{+\infty} \langle u_{mn}(R,\psi,t)u_{mn}(R,\psi,t+\tau)\rangle e^{-i\omega\tau}d\tau = \left|H_{p_{ff}}(m,n,R,\psi,\omega)\right|^2 C_{u_{mn}u_{mn}} \tag{12.41}$$

where

$$C_{v_{mn}v_{mn}} = \frac{4}{\pi^2\ell^2}\int_0^{\ell}\int_0^{\ell}\int_0^{2\pi}\int_0^{2\pi} S_f(\xi_1,\eta_1,\xi_2,\eta_2,\omega)\sin\frac{m\pi\xi_1}{\ell}$$
$$\sin\frac{m\pi\xi_2}{\ell}\cos n\eta_1\cos n\eta_2 d\xi_1 d\eta_1 d\xi_1 d\eta_2 \tag{12.42}$$

$$C_{u_{mn}u_{mn}} = \frac{4}{\pi^2\ell^2}\int_0^{\ell}\int_0^{\ell}\int_0^{2\pi}\int_0^{2\pi} S_f(\xi_1,\eta_1,\xi_2,\eta_2,\omega)\sin\frac{m\pi\xi_1}{\ell}$$
$$\sin\frac{m\pi\xi_2}{\ell}\sin n\eta_1\sin n\eta_2 d\xi_1 d\eta_1 d\xi_1 d\eta_2 \tag{12.43}$$

Thus,

$$[\tilde{p}_{ff}(R,\psi,\omega)]_{Ave.\varphi} = \frac{1}{2}\sum_{m=1}^{\infty}\sum_{n=0}^{\infty}\left|H_{p_{ff}}(R,\psi,m,n,\omega)\right|^2 \bar{C}_{mn} \tag{12.44}$$

where

$$\bar{C}_{mn} = \frac{4}{\pi^2\ell^2}\int_0^{\ell}\int_0^{\ell}\int_0^{2\pi}\int_0^{2\pi} S_f(\xi_1,\eta_1,\xi_2,\eta_2,\omega)\sin\frac{m\pi\xi_1}{\ell}$$
$$\sin\frac{m\pi\xi_2}{\ell}\cos n(\eta_1-\eta_2)d\xi_1 d\eta_1 d\xi_2 d\eta_2 \tag{12.45}$$

12.3.2 Far Field due to Turbulence Excitation

The cross-spectrum function for turbulent pressure on the cylinder can be written as follows[5,6]:

$$S_f(\xi_1,\eta_1,\xi_2,\eta_2,\omega) = \tilde{p}(\omega)e^{-\frac{k}{c_0}|\xi_1-\xi_2|}\cos k(\xi_1-\xi_2)e^{-\beta|\eta_1-\eta_2|} \tag{12.46}$$

where $\tilde{p}(\omega)$ is the spectrum of the turbulent pressure and

$$k = \frac{\omega}{U_c},\ \beta = .7\frac{\omega a}{U_c},\ C_o = \frac{1}{.12},\ U_c = .65\,U_o \tag{12.47}$$

in which ω is the radian frequency, U_c is the convection velocity (which is 0.65 U_o, U_o being the free stream flow velocity), and a is the cylinder radius.

Substituting the cross spectrum of the turbulence pressure given by Equation 12.46 into $\bar{C}_{mn}$ and dividing $\bar{C}_{mn}$ into two parts, the result is

$$\bar{C}_{mn} = \bar{C}_1 \quad \bar{C}_2 \tag{12.48}$$

where

$$\bar{C}_1 = \frac{1}{\ell^2}\int_0^{\ell}\int_0^{\ell} e^{-\frac{k}{c_o}|\xi_1-\xi_2|}\cos k(\xi_1-\xi_2)\sin\frac{m\pi\xi_1}{\ell}\sin\frac{m\pi\xi_2}{\ell}\,d\xi_1 d\xi_2 \tag{12.49}$$

and

$$\bar{C}_2 = \frac{4}{\pi^2}\int_0^{2\pi}\int_0^{2\pi} e^{-\beta|\eta_1-\eta_2|}\cos n(\eta_1-\eta_2)\,d\eta_1 d\eta_2 \tag{12.50}$$

The value of $\bar{C}_1$ can be obtained directly from Bozich[7] and is

$$\begin{aligned}
\bar{C}_1 = \Bigg[\frac{e^{-T\ell}}{2}\Bigg\{\Bigg[&\frac{P\ell}{(T\ell)^2+(P\ell)^2}+\frac{Q\ell}{(T\ell)^2+(Q\ell)^2}\Bigg]\\
&\Bigg[\frac{\{-T\ell\sin P\ell - P\ell\cos P\ell\}}{(T\ell)^2+(P\ell)^2}+\frac{\{-T\ell\sin Q\ell - Q\ell\cos Q\ell\}}{(T\ell)^2+(Q\ell)^2}\Bigg]\\
&-\Bigg[\frac{T\ell}{(T\ell)^2+(Q\ell)^2}-\frac{T\ell}{(T\ell)^2+(P\ell)^2}\Bigg]\\
&\Bigg[\frac{\{Q\ell\sin Q\ell - T\ell\cos Q\ell\}}{(T\ell)^2+(Q\ell)^2}-\frac{\{P\ell\sin P\ell - T\ell\cos P\ell}{(T\ell)^2+(P\ell)^2}\Bigg]\Bigg\}\\
&+\frac{1}{2}\Bigg[\frac{T\ell}{(T\ell)^2+(Q\ell)^2}+\frac{T\ell}{(T\ell)^2+(P\ell)^2}\Bigg]\\
&+\frac{1}{2}\Bigg[\frac{Q\ell}{(T\ell)^2+(Q\ell)^2}+\frac{P\ell}{(T\ell)^2+(P\ell)^2}\Bigg]^2\\
&-\frac{1}{2}\Bigg[\frac{T\ell}{(T\ell)^2+(Q\ell)^2}-\frac{T\ell}{(T\ell)^2+(P\ell)^2}\Bigg]^2\Bigg]
\end{aligned} \tag{12.51}$$

where

$$T\ell = \frac{k\ell}{C_o}, \quad P\ell = m\pi + k\ell, \quad Q\ell = m\pi - k\ell \tag{12.52}$$

The value of $\bar{C}_2$ is computed as follows:

$$\begin{aligned}\frac{\bar{C}_2}{4/\pi^2} &= 2\int_0^{\pi} d\eta_2 \int_0^{\eta_2} e^{-\beta(\eta_2-\eta_1)} \cos n(\eta_2 - \eta_1) d\eta_1 \\ &+ 2\int_0^{\pi} d\eta_2 \int_{\eta_2}^{2\pi} e^{-\beta(\eta_1-\eta_2)} \cos n(\eta_1 - \eta_2) d\eta_1\end{aligned} \tag{12.53}$$

$$\begin{aligned}&= 2\int_0^{\pi} d\eta_2 \left[\frac{e^{-\beta\eta_1}(-\beta\cos n\eta_1 + n\sin n\eta_1)}{\beta^2 - n^2}\right]_0^{\eta_2} \\ &+ 2\int_0^{\pi} d\eta_2 \left[\frac{e^{-\beta\eta_1}(-\beta\cos n\eta_1 + n\sin n\eta_1)}{\beta^2 + n^2}\right]_0^{\pi-\eta_2}\end{aligned} \tag{12.54}$$

Finally,

$$\bar{C}_2 = \frac{16}{\pi^2(\beta^2 + n^2)}\left\{\frac{(\beta^2 - n^2)(e^{-\beta\pi}\cos n\pi - 1)}{\beta^2 + n^2} + \beta\pi\right\} \tag{12.55}$$

12.4 Computer Programs

The computer program for the equations in the previous section is obtained by some small revisions of the program for point loading of cylindrical shells. The basic program for calculation of the frequency response function for far field pressure $H_{p_{ff}}$ remains the same, but is now coupled with the "joint acceptance" function given by $\bar{C}_{mn}$. A listing of the program is given in Table 12.1. The input parameters are very similar to the ones for point loading of cylindrical shells as given in Chapter 7.

The input parameters for Table 12.1 are listed in statements 50–110 and the corresponding numbers are given in statements 3202–3214. In the random

TABLE 12.1

Computer Program for Random Loading of Cylindrical Shells

```
RANDOMPR. BAS
10      DIM E(50, 50), F(50, 50)
30      DIM J(50), T(50), Y(50), L(50), H(50), K(50), N(50)
50      READ C5, C2, C1, C4, R1, C3, M, P1, P2, P3, D, K9, L9
70      READ S7, S8, S9, Q, Q9, B8, W8, N, N2
80      READ B2, B3, B4, B5, B6, B7
90      READ M2, M3, M4, M5, M6, M7, M8, M9
100     READ E4, E5, E6, E7, E8, E9
110     READ F4, F5, F6, F7, F8, F9
120     FOR I = 0 TO Q
130     IF I > 0 THEN 160
140     LET N (0) = .5
150     GOTO 170
160     LET N (I) = 1
170     NEXT I
180     FOR P = P1 TO P2 STEP P3
190     LET C6 = P*3.14159*C4
210     IF C3 > C6 THEN 3215
220     LET Z = SQR((C6^2)-(C3^2))
230     GOSUB 3460
240     FOR I = 0 TO N2
250     IF C3 > C6 THEN 870
270     LET U4 = 0
280     LET A = Z*K(I + 1)-I*K(I)
290     LET X4 = (C3*K (I)) /A
300     GOTO 980
870     LET B = (I/Z) *J (I) - J (I + 1)
880     LET C = (I/Z) *Y (I)-Y (I+1)
890     IF ABS (J(I) / (10^K9)) < ABS (Q9 *Y (I)) THEN 950
900     LET D2 = Z* ((B^2) + (10^ (2*K9)) * (C^2))
910     LET U4 = (C3* (10^K9) * (J(I) *C-Y (I) *B)) /D2
920     LET 07 = J (I) *B
930     LET X4 = - (C3* (07 + (10^(2*K9)) *Y (I) *C)) /D2
940     GOTO 980
950     LET U4 = 0
960     LET X4 = -C3 *Y (I) / (I*Y (I) - Z*Y (I+1))
980     GOSUB 5000
1450    NEXT I
1460    NEXT P
2600    FOR S6 = S7 TO S8 STEP S9
2610    LET G = SIN (S6)
2620    LET G1 = C3* COS (S6)
2630    LET Z = C3* G
```

TABLE 12.1 (Continued)

Computer Program for Random Loading of Cylindrical Shells

```
2635    LET M = N
2640    GOSUB 3220
2670    FOR P = P1 TO P2 STEP P3
2680    LET U7 = SIN (P*1.5708) *COS (1.5708* (P-1))
2690    LET V7 = COS (P*1.5708) *SIN (1.5708* (P+1))
2700    LET C6 = P*3.1416 *C4
2710    LET G2 = (2/G) * (1/ (1 - (G1/C6)^2))
2720    LET G4 = ((2/C6) *SIN (G1/ (2*C4))) * (1/6.2832)
2730    LET G3 = ((2/C6) *COS (G1/ (2*C4))) * (1/6.2832)
2740    LET U9 = U7 *G2 *G3
2750    LET V9 = V7 *G2 *G4
2751    T = .12*W8
2752    P8 = P*3.1416 + W8
2753    Q8 = P*3.1416 - W8
2754    D8 = (T^2) + (P8^2)
2755    D9 = (T^2) + (Q8^2)
2756    U1 = SIN (P8)
2757    U2 = COS (P8)
2758    E1 = SIN (Q8)
2759    E2 = COS (Q8)
2760    D1 = P8/D8
2761    E3 = Q8/D9
2762    X1 = D1 + E3
2763    X2 = ((- T*U1 - P8*U2) /D8) + ((-T*E1 - Q8*E2) /D9)
2764    X3 = T/D9
2765    X4 = T/D8
2766    X5 = X3 - X4
2767    X6 = ((Q8*E1 - T*E2) /D9) - ((P8*U1 - T*U2) /D8)
2768    X7 = .5* (X3 + X4)
2769    X8 = .5* ((E3 + D1) ^2)
2770    X0 = .5* ((X3 - X4) ^2)
2771    Z9 = .5* EXP (- T) * (X1*X2 - X5*X6) + .5*(X7 + X8 - X0)
2779    FOR I = 0 TO N2
2780    LET E9 = COS (1.5708*I)
2790    LET F9 = SIN (1.5708*1)
2800    LET G7 = (I/Z) *J (I) - J (I+1)
2810    LET G8 = (I/Z) *Y (I) - Y (I+1)
2820    LET J1 = E (P - P1 + 1, I) *E9 + F (P - P1 + 1, I) *F9
2830    LET J2 = F (P - P1 + 1, I) *B9 - E (P - P1 + 1, I) *F9
2840    IF ABS (J (I) / (10^K9)) <ABS (Q9 *Y (I)) THEN 2910
2850    LET G9 = (G7^2) + (10^ (2*K9)) * (G8^2)
2860    LET J3 = U9*G7 + V9*(10^K9)*G8
```

(Continued)

TABLE 12.1 (Continued)

Computer Program for Random Loading of Cylindrical Shells

```
2870    LET J4 = U9* (10^K9) *G8 - V9 *G7
2880    LET J5 = (J1 * J3 - J2 *J4) /G9
2890    LET J6 = (J2* J3 + J1 *J4) /G9
2900    GOTO 2930
2910    LET J5 = (J1 *V9 - J2 *U9) / ((10^K9) *G8)
2913    IF ABS(J5) > (10^(- 15)) THEN 2920
2915    J5 = 0
2920    LET J6 = (J2 *V9 + J1 *U9) / ((10^K9) *G8)
2923    IF ABS (J6) > (10^ (- 15)) THEN 2930
2927    J6 = 0
2930    LET P4 = C2* (C1^2) *C3
2932    P5 = ((J5*P4) ^2) + ((J6*P4) ^2)
2933    Y1 = (B8^2) + (I^2)
2934    Y2 = (B8^2) - (I^2)
2935    Y3 = 16/ ((3.1416^2) *Y1)
2936    Y4 = B8* 3.1416
2937    Y5 = EXP (- Y4) *COS (I*3.1416)
2940    X9 = Y3* (((Y2* (Y5 - 1)) /Y1) + Y4)
2950    Y9 = X9* Z9* (N (I) ^2)
2960    W5 = W5 + P5* Y9
3010    NEXT I
3020    NEXT P
3030    W7 = SQR (W5)
3065    X = W7 *COS (S6)
3067    Y = W7 *SIN (S6)
3070    LPRINT "S6 ="; S6; "W5 =";W5; "W7 ="; W7
3092    LPRINT "X ="; X; "Y ="; Y
3100    LET W5 = 0
3200    NEXT S6
3201    STOP
3202    DATA .05, 41.5, .29, .16, .13, 3, 18, 1, 14, 1, .3, 0, 0
3206    DATA .1, 1.6, .1, 12, .00000001, 636, 5700, 16, 12
3208    DATA 0, 0, 1, 1, -1, 1
3210    DATA 1, 1, 1, -1, 2, 1, 1, 1
3212    DATA 1, .3, 0, .000003332,0, .35
3214    DATA .00000116, 1, 0, .000003332, .00000116, .00000100
3215    LET Z = SQR ((C3^2) - (C6^2))
3216    GOSUB 3220
3217    GOTO 240
3220    LET T (M) = 0
3230    LET T (M - 1) = 10^(-L9)
3240    FOR I = M TO 2 STEP -1
3250    LET T (I - 2) = 2* (I - 1) * (1/Z) *T (I - 1) -T (I)
```

TABLE 12.1 (Continued)

Computer Program for Random Loading of Cylindrical Shells

```
3260    NEXT I
3270    LET S = 0
3280    FOR I = 1 TO M/2
3290    LET S = S+2 *T (2*I)
3300    NEXT I
3310    LET S1 = S + T (0)
3320    FOR I = 0 TO M
3330    LET J (I) = (T (I)) /S1
3340    NEXT I
3350    LET S2 = 0
3360    FOR I = 1 TO M/2
3370    LET S2 = S2 + (2/I) *COS ((I-1) *3.1416) *J (2*I)
3380    NEXT I
3390    LET Y (0) = ((2/3.1416) * (J (0)* (LOG (.5*Z) +
        .577215665#) +S2)) /(10^K9)
3400    LET O9 = Y(0) * (10^K9)
3410    LET Y(1) = ((1/J (0)) * (J(1) *O9 - (2/ (3.1416*Z))))/
        (10^K9)
3420    FOR I = 1 TO M
3430    LET Y (I + 1) = 2* I* (1/Z) *Y (I) - Y (I - 1)
3440    NEXT I
3450    RETURN
3460    LET L (M) = 0
3470    LET L (M - 1) = 10^(-L9)
3480    FOR I = M TO 2 STEP - 1
3490    LET L (I - 2) = 2* (I - 1) * (1/Z) *L (I - 1) + L (I)
3500    NEXT I
3510    LET S3 = 0
3520    FOR I = 1 TO M
3530    LET S3 = S3 + 2*L (I)
3540    NEXT I
3550    LET S4 = EXP(- Z) * (S3 + L (0))
3560    FOR I = 0 TO M
3570    LET H (I) = (L (I))/S4
3580    NEXT I
3590    IF Z < 2 THEN 3610
3600    IF Z>2 THEN 3660
3610    LET U = .5*Z
3620    LET A = -LOG (.5*Z) *H (0) - .57721566# + .4227842# *U^2
3630    LET B = A + .23069756# *U^4 + .0348859 *U^6 + 2.62698E
        - 03 *U^8
3640    LET K (0) = (B + .0001075 *U^10 + .0000074 *U^12) / (10^K9)
3650    GOTO 3710
```

(Continued)

TABLE 12.1 (Continued)

Computer Program for Random Loading of Cylindrical Shells

```
3660  LET U = .5/Z
3670  LET V = (Z^ .5) *EXP (Z)
3680  LET A = 1.25331414# - 7.832358E - 02*U + 2.189568E
      - 02* U^2
3690  LET B = A - 1.062446E - 02 *U^3 + 5.87872E - 03* U^4
3700  LET K (0) = ((B- .0025154 *U^5 + 5.3208E - 04* U^6) /V)
      / (10^K9)
3710  LET O8 = K (0) * (10^K9)
3720  LET K (1) = ((1/H(0)) * ((1/Z) - O8*H (1))) / (10^K9)
3730  FOR I = 1 TO M
3740  LET K (I + 1) = 2*I* (1/Z) *K (I)+K(I - 1)
3750  NEXT I
3760  RETURN
5000  LET A0 = (I^2) * (.5* (1 - D) *E8 - E9 - M9* B4* F4)
      - E4* (C6^2) + ((C3*C1)^2)
5005  LET K0 = A0 - (C3*C1) ^ 2
5010  LET A1 = I*C6 * (E5 + .5* (1 + D)*B8 + E9)
5015  LET A2 = C6* (E5 + (C6 ^ 2) * (E6 + M7*E7) + (I ^ 2) *
      (E8 - M9*F4))
5020  LET A3 = A1
5025  LET N4 = (I ^ 2) * (-F5 - F6) + (C6^2) * (-1.5* (1
      - D)*E8 - E9 + (M6*M5 - M8)*F8)
5030  LET A4 = N4 + ((C3*C1)^2)
5035  LET N5 = I* (-F5 + (B3 - 1)*F6 + (B5 - M2)*F7) +
      (I^3)*(-F6 + (M2 - 1)*F7)
5040  LET A5 = N5 + I* (C6^2) * (-E8 - (M8 + M6)*F8 - F9)
5045  LET A6 = C6* (E5 + (C6^2) * (E6*B2 - B7*B6) + (I^2)*C6*
      (E8 - M3*F4))
5050  LET K4 = N4
5055  LET N7 = I* (-F5) + (I^3) * (-F6) + I* (C6^2) *
      (-B7*F9 - E8 + M6*M5*F8 - M4*F4)
5060  LET A7 = N7
5065  LET N8 = (I^2) * ((B5 + M6)*F7 - 2*F6) - (I^4)*F7
      - (C6^4)*E7 - 2* (C6^2)*D*E8
5070  LET I8 = N8 + (I^2) * (C6^2) * (-2*F9 - M6*F8 - 2*F4)
      - F7*M2 + B3*F6
5075  A8 = I8 - F5 + ((C3*C1) ^2) + C2*C3* (C1^2)*X4
5080  LET I3 = A0* (A4*A8 - A5*A7) - A1* (A3*A8 - A5*A6)
5085  LET R3 = I3 + A2* (A3*A7 - A4*A6)
5090  LET R4 = A0*A4 - A1*A3
5095  LET K8 = I8 - F5
5100  LET T1 = K0*K4 - A1*A3
5105  LET T2 = K0*A7 - A1*A6
5110  LET T3 = A3*A7 - K4*A6
```

TABLE 12.1 (Continued)

Computer Program for Random Loading of Cylindrical Shells

```
5115    LET T4 = (A5*T2 - A2*T3)/T1
5120    LET T5 = T4 - K8
5125    LET T6 = SQR (T5)
5130    LET R8 = C2*C3* (C1^2)*u4 + 2*C5*C3*C1*T6
5135    LET R5 = (R3^2) + (R4*R8) ^2
5140    IF ABS(R3/R4) >R8 THEN 5160
5145    PRINT "RES"; "I ="; I; "P ="; P
5150    LET F(P - P1 + 1, I) = (R4^2)*R8/R5
5155    GOTO 5165
5160    LET F(P - P1 + 1, I} = (R4^2)*R8/R5
5165    LET E(P - P1 + 1, I) = R4*R3/R5
5170    RETURN
7000    END
```

program shown in Table 12.1, the input parameters are also given in statements 50–110 and the corresponding numbers in statements 3202–3214. In the random program, L8 (the length) and Q5, Q6, Q7 (the location of the point load) are omitted; B8 and W8 are added in statement 70, with the corresponding numbers given in statement 3206. In the random program shown in Table 12.1 (statement 70),

$$B8 = 1.077\frac{\omega a}{U_o}$$
$$W8 = 1.54\frac{\omega \ell}{U_o} \tag{12.56}$$

where
ω = radian frequency
U_o = free stream flow velocity
a = shell radius
ℓ = shell length

All other parameters are the same as in the point loading program. A sample output from the program is given in Table 12.2 corresponding to the input parameters given in the listing of Table 12.1.*

For cases in which we are considering the response of stiffened shells to turbulence, the structure between the stiffeners is vibrating and radiating. If we assume nodal points at the stiffeners, then the shell plating between the stiffeners constitutes a short shell. In this situation, a number of simplifications

* Some numerical output is shown in Table 12.2.

TABLE 12.2

Some Numerical Output for the Computer Program Shown in Table 12.1

```
S6= S6 = .1 W5 = 4.164132E - 07 W7 = 6.453009E - 04
X= 6.42077E - 04 Y = 6.442259E - 05
S6= .2 W5 = 7.047662E - 07 W7 = 8.395035E - 04
X= 8.227694E - 04 Y = 1.667836E - 04
S6= .3 W5 = 8.885333E - 07 W7 = 9.426204E - 04
X= 9.005198E - 04 Y = 2.785634E - 04
S6= .4 W5 = 9.77337E - 07 W7 = 9.886036E - 04
X= 9.105642E - 04 Y = 3.849804E - 04
S6= .5 W5 - 1.093021E - 06 W7 = 1.045476E - 03
X= 9.174916E - 04 Y = 5.01228E - 04
S6= .6 W5 = 1.210116E - 06 W7 = 1.100053E - 03
X= 9.079126E - 04 Y = 6.211365E - 04
S6= .7000001 W5= 1.311023E - 06 W7= 1.144999E - 03
X= 8.757436E - 04 Y= 7.376288E - 04
S6= .8000001 W5= 1.423244E - 06 W7= 1.192998E - 03
X= 8.311696E - 04 Y= 8.558044E - 04
S6= .9000001 W5= 1.571214E - 06 W7= 1.253481E - 03
X= 7.791761E - 04 Y= 9.818854E - 04
S6= 1 W5= 1.737153E - 06 W7= 1.318011E - 03
X= 7.121242E - 04 Y= 1.109068E - 03
S6= 1.1 W5= 1.863715E - 06 W7= 1.365179E - 03
X= 6.1924E - 04 Y= 1.216658E - 03
S6= 1.2 W5= 1.951203E - 06 W7= 1.396855E - 03
X= 5.061609E - 04 Y= 1.301923E - 03
S6= 1.3 W5= 2.053501E - 06 W7= 1.433004E - 03
X= 3.833267E - 04 Y= 1.380783E - 03
S6= 1.4 W5= 2.136107E - 06 W7= 1.461543E - 03
X= 2.484139E - 04 Y= 1.440277E - 03
S6= 1.5 W5= 2.139173E - 06 W7= 1.462591E - 03
X= 1.034592E - 04 Y= 1.4589
```

can be made to the program. For $m\pi a/1 \gg \omega a/c_o$, $\beta \gg n$ and $k\ell \gg m\pi$, the program simplifies considerably. A second version of the program is given in Table 12.3, which employs, these assumptions.

The corresponding output is given in Table 12.4. In these cases, the frequencies in water can be computed explicitly because of the simple form taken by the reactive impedance (virtual mass function). A series of horizontal directivity patterns (using peripheral average as shown in the analysis) for the far field radiation from turbulence excitation is shown in Figures 12.2–12.5. The shell for which these calculations were made had the same parameters as the shell in Chapter 7. The radiation described here, however, is that due to the vibration of the shell plating between stiffeners.

TABLE 12.3

Simplified Computer Program of Table 12.1

```
PROG 4.BAS
10      DIM E(50, 52), F(50, 52)
30      DIMJ(52), T(52), Y(52), L(52), H(52), K(52), N(52)
50      READ C5, C2,, C1, C4, R1, C3, M, P1, P2, P3, D, K9, L9
70      READ S7, S8, S9, Q, Q9, B8, W8, N, N2
80      READ B2, B3, B4, B5, B6, B7
90      READ M2, M3, M4, M5, M6, M7, M8, M9
100     READ E4, E5, E6, E7, E8, E9
110     READ F4, F5, F6, F7, F8, F9
120     FOR I = Q TO N2
130     IF I > 0 THEN 160
140     LET N(0) = .5
150     GO TO 170
160     LET N(I) = 1
170     NEXT I
180     FOR P = P1 TO P2 STEP P3
185     FOR I = Q TO N2
190     LET C6 = P*3.1416*C4
192     LET Z = C6
195     GO TO 292
210     IF C3 > C6 THEN 3215
220     LET Z = SQR((C6^2) - (C3^2))
230     GOSUB 3460
250     IF C3 > C6 THEN 870
270     LET U4 = 0
280     LET A = Z*K(I + 1) - I*K(I)
290     LET X4 = (C3*K(I))/A
292     IF I > 2 THEN 300
294     LET X4 = C3* (Z - I)^(- 1)
296     GO TO 980
300     LET H2 = LOG((I/Z) + SQR(I + (I/Z)^2))
305     LET H3 = LOG(((I + 1)/Z) + SQR(I + ((I + 1)/Z)^2))
310     LET H1 = LOG(((I - 1)/Z) + SQR(I + ((I - 1)/Z)^2))
315     LET L2 = SQR(1+(Z/I)^2)
320     LET L3 = SQR(1 + (Z/(I - 1))^2)
325     LET L1 = SQR(1 + (Z/(I - 1))^2)
330     LET L4 = 1/L2
335     LET L5 = 1/L3
340     LET L6 = 1/L1
345     LET H4 = EXP(I* (L2-H2) - ((I - 1) * (L1 - H1)))
350     LET H5 = EXP(I* (L2-H2) - ((I + 1) * (L3 - H3)))
355     LET H6 = SQR((L6*I)/(L4* (I - 1)))
360     LET H7 = SQR ((L5*I)/(L4* (1+I)))
```

(Continued)

TABLE 12.3 (Continued)

Simplified Computer Program of Table 12.1

```
365    LET X4 = (2*C3/Z) * (1/(H4*H6 + H5*H7))
370    GO TO 980
870    LET B = (I/Z)*J(I) - J(1 + 1)
880    LET C = (I/Z)*Y(I) - Y(I + 1)
890    IF ABS (J(I)/(10^K9))<ABS(Q9*Y(I)) THEN 950
900    LET D2 = Z* ((B^2)+ (10^(2*K9)) + (C^2))
910    LET U4 = (C3* (10^K9) * (J(I)*C - Y(I)*B))/D2
920    LET 07 = J(I)*B
930    LET X4 = - (C3* (07 + (10^(2*K9))*Y(I) + C))/D2
940    GO TO 980
950    LET U4 = 0
960    LET X4 = - C3*Y(I)/(I*Y(I) - Z*Y(I + 1))
980    GOSUB 5000
1450   NEXT I
1460   NEXT P
2600   FOR S6 = S7 TO S8 STEP S9
2610   LET G = SIN(S6)
2620   LET G1 = C3*COS(S6)
2630   LET Z = C3*G
2635   LET M = N
2640   GOSUB 3220
2670   FOR P = P1 TO P2 STEP P3
2680   LET U7 = SIN(P* 1.5708)*COS(1.5708* (P - 1))
2690   LET V7 = COS(P*1.5708)*SIN(1.5708* (P + 1))
2700   LET C6 = P*3.1416*C4
2710   LET G2 = (2/G) * (1/(1 - (G1/C6)^2))
2720   LET G4 = [(2/C6)*SIN(G1/(2*C4))] * (1/6.2832)
2730   LET G3 = [(2/C6) *COS(G1/(2*C4))] * (1/6.2832)
2740   LET U9 = U7*G2*G3
2750   LET V9 = V7*G2*G4
2771   LET Z9 = .12/W8
2779   FOR I = Q TO N2
2780   LET Q6=COS(1.5708*I)
2790   LET Q7 = SIN(1.5708*I)
2800   LET G7 = (I/Z)*J(I) - J(I + 1)
2810   LET G8 = (I/Z)*Y(I) - Y(I+1)
2820   LET J1 = E(P - P1 + 1, I)*Q6 + F(P - P1 + 1, I)*Q7
2830   LET J2 = F(P - P1 + 1, I)*Q6 - E(P - P1 + 1, I)*Q7
2840   IF ABS(J(I)/(10^K9))<ABS(Q9*Y(I)) THEN 2910
2850   LET G9 = (G7^2) + (10^(2*K9)) * (G8^2)
2860   LET J3 = U9*G7 + V9* (10^K9)*G8
2870   LET J4 = U9*(10^K9)*G8 - V9*G7
2880   LET J5 = (J1*J3 - J2*J4)/G9
```

TABLE 12.3 (Continued)

Simplified Computer Program of Table 12.1

```
2890    LET J6 = (J2*J3 + J1*J4)/G9
2900    GO TO 2930
2910    LET J5 = (J1*V9 - J2*09)/((10^K9)*G8)
2913    IF ABS(J5) > (10^(- 15)) THEN 2920
2915    LET J5 = 0
2920    LET J6 = (J2*V9 + J1*U9)/((10^K9)*G8)
2923    IF ABS(J6) > (10^(- 15)) THEN 2930
2927    LET J6 = 0
2930    LET P4 = C2* (C1^2)*C3
2932    LET P5 = ((J5*P4)^2) + ((J6*P4)^2)
2940    LET X9 = 5.09/B8
2950    LET Y9 = X9*Z9* (N(I)^2)
2960    LET W5 = W5 + P5*Y9
3010    NEXT I
3020    NEXT P
3030    W7 = SQR(W5)
3065    LET X = W7*SIN(S6)
3067    LET Y = W7*COS(S6)
3070    PRINT "S6 =";S6; "W5 ="; W5; "W7 ="; W7
3092    PRINT "X =";X; "Y =";Y
3100    LET W5 = 0
3800    NEXT S6
3201    STOP
3202    DATA .05, 41.5, .29, 4, .13, 4, 2, 1, 5, 1, .3, 0, 0
3206    DATA .4, 1.6, .2, 0, .00000001, 848, 7600, 28, 27
3208    DATA 0, 0, 1, 1, - 1, 1
3210    DATA 1, 1, 1, - 1, 2, 1, 1, 1
3212    DATA 1, .3, 0, .000000833, 0, .35
3214    DATA .00000029, 1, 0, .000000833, .00000029, .00000025
3215    LET Z = SQB((C3^2) - (C6^2))
3216    GO SUB 3220
3217    GO TO 185
3220    LET T(M) = 0
3230    LET T(M - 1) = 10^(- L9)
3240    FOR I = M TO 2 STEP - 1
3250    LET T(I - 2) = 2* (I - 1) * (1/Z)*T(I - 1) - T(I)
3260    NEXT I
3270    LET S = 0
3280    FOR I = 1 TO M/2
3290    LET S = S + 2*T(2*I)
3300    NEXT I
3310    LET S1 = S + T(0)
```

(Continued)

TABLE 12.3 (Continued)

Simplified Computer Program of Table 12.1

```
3320   FOR I = 0 TO M
3330   LET J(I) = (T(I))/SI
3340   NEXT I
3350   LET S2 = 0
3360   FOR I = 1 TO M/2
3370   LET S2 = S2 + (2/I)*COS((I - 1)*3.1416)*J(2*I)
3380   NEXT I
3390   LET Y(0) = ((2/3.1416) * (J(0) * (LOG(.5*Z) +
       .577215665) + S2))/(10^K9)
3400   LET 09 = Y(0) * (10^K9)
3410   LET Y(1) = ((I/J(0)) * (J(1)*09 - (2/(3.1416*Z))))/
       (10^K9)
3420   FOR I = 1 TO M
3430   LET Y(I + 1) = 2*I* (I/Z)*Y(I) - Y(I - 1)
3440   NEXT I
3450   RETURN
3460   LET L(M) = 0
3470   LET L(M - 1) = 10^(- L9)
3480   FOR I = M TO 2 STEP - 1
3490   LET L(I - 2) = 2* (I - 1) * (1/Z) * L(I - 1) + L(I)
3500   NEXT I
3510   LET S3 = 0
3520   FOR I = 1 TO M
3530   LET S3 = S3 + 2*L(I)
3540   NEXT I
3550   LET S4 = EXP(- Z) * (S3 + L(0))
3560   FOR I = 0 TO M
3570   LET H(I) = (L(I))/S4
3580   NEXT I
3590   IF Z < 2 THEN 3610
3600   IF Z > 2 THEN 3660
3610   LET U = .5*Z
3620   LET A = - LOG(.5*Z)*H(0) - .57721566 + .42278420*0^2
3630   LET B = A + .23069756*U^4 + .03488590*U^6 + .00262698*U^8
3640   LET K(0) = (B + .00010750*U^10 + .00000740*U^12)/(10^K9)
3650   GO TO 3710
3660   LET U = .5/Z
3670   LET V = (Z^.5)*EXP(Z)
3680   LET A = 1.25331414 - .07832358*U + .02189568*U^2
3690   LET B = A - .01062446*U^3 + .00587872*U^4
3700   LET K(0) = ((B -.00251540*U^5 + .00053208*U^6)/V)/(10^K9)
3710   LET 08 = K(0) * (10^K9)
3720   LET K(1) = ((1/H(U)) * ((1/Z) - 08*H(1)))/(10^K9)
```

TABLE 12.3 (Continued)

Simplified Computer Program of Table 12.1

```
3730   FOR I = 1 TO M
3740   LET K(I + 1) = 2*I* (I/Z)*K(I) +K(I - 1)
3750   NEXT I
3760   RETURN
4010
5000   LET A0 = (I^2) * (.5* (1 - D)*E8 - E9 - M9*B4* F4)
       - E4* (C6^2) + ((C3*C1)^2)
5005   LET KU = AU - (C3*C1)^2
5010   LET A1 = I*C6* (E5 + .5* (1 + D)*E8 + E9)
5015   LET A2 = C6* (E5 + (C6^2) * (E6 + M7*E7) + (I^2) *
       (E8-M9*F4))
5020   LET A3 = A1
5025   LET N4 = (I^2) * (- F5 - F6) + (C6^2) * (- 1.5* (1
       - D)*E8 - E9+(M6*M5 - M8)*F8)
5030   LET A4 = N4 + ((C3*C1)^2)
5035   LET N5 = I * (- F5 + (B3 - 1) *F6 + (B5 - M2)*F7 +
       (I^3) * (- F6 + (M2 - 1)*F7)
5040   LET A5 = N5 + I * (C6^2) * (- E8 - (M8 + M6)*F8 - F9)
5045   LET A6 = C6* (E5 + (C6^2) * (E6*B2 - E7*B6)+(I^2) * (E8
       - M3*F4))
5050   LET K4 = N4
5055   LET N7 = I * (- F5) + (I^3) * (- F6) + I * (C6^2) *
       (- B7*F9 - E8 + M6*M5*F8 - M4*F4)
5060   LET A7 = N7
5065   LET N8 = (I^2) * ((B5 + M6)*F7 - 2*F6) - (I^4)*F7
       - C6^4) * E7 - 2* (C6^2)*D* E8
5070   LET I8 = N8 + (I^2) * (C6^2) * (-2*F9 - M6*F8 - 2*F4)
       - F7*M2 + B3*F6
5075   LET A3 = I8 - F5 + ((C3*C1)^2) + C2*C3* (C1^2)*X4
5080   LET I3 = A0 * (A4*A8 - A5*A7) - A1 * (A3*A8 - A5*A6)
5085   LET R3 = I3 + A2* (A3*A7 - A4*A6)
5090   LET R4 = A0*A4 - A1*A3
5095   LET K8=I8 - F5
5100   LET T1 = K0*K4 - A1*A3
5105   LET T2 = K0*A7 - A1*A6
5110   LET T3 = A3*A7 - K4*A6
5115   LET T4 = (A5*T2 - A2*T3)/T1
5120   LET T5 = T4 - K8
5122   LET T6 = SQR(T5)
5124   LET W = T6/CI
5125   LET F = SGR(1+C2*X4/C3)
5126   LET W1 = W/F
5130   LET R8 = C2*C3* (C1^2)*U4 + 2*C5*C3*C1*T6
```

(*Continued*)

TABLE 12.3 (Continued)

Simplified Computer Program of Table 12.1

```
5135   LET R5 = (R3^2) + (R4*R8)^2
5140   IF ABS (R3/R4)>R8 THEN 5160
5145   PRINT "RES"; "I =";I; "P =";P
5150   LET F(P - P1 + 1, I) = (R4^2)*R8/R5
5155   GO TO 5165
5160   LET F(P - P1 + 1, I) = (R4^2)*R8/R5
5165   LET E(P - P1 + 1, I) = R4*R3/R5
5170   RETURN
7000   END
```

TABLE 12.4

Some Numerical Output for the Computer Program Shown in Table 12.3

```
RESI = 5 P = 3
RESI = 6 P = 3
RESI = 7 P = 3
RESI = 8 P = 3
RESI = 9 P = 3
RESI = 10 P = 3
RESI = 11 P = 3
RESI = 12 P = 3
RESI = 13 P = 3
RESI = 14 P = 3
S6 = 0.4 W5 = 1.21352E - 6 W7 = 1.1016E - 3
K = 4.28983E - 4 Y = 1.01464E - 3
S6 = 0.6 W5 = 1.4497F - 6 W7 = 1.20404E - 3
X = 6.79849E - 4 Y = 9.93733E - 4
S6 = 0.8 W5 = 1.73743E - 6 W7 = 1.31311E - 3
X = 9.45557E - 4 Y = 9.18339E - 4
S6 = 1. W5 = 1.96313E - 6 W7 = 1.40112E - 3
X = 0.001179 Y = 7.57027E - 4
S6 = 1.2 W5 = 2.13326E - 6 W7 = 1.46057E - 3
X = 1.36131E - 3 Y = 5.29243E - 4
S6 = 1.4 W5 = 2.23284E-6 W7 = 1.49427E - 3
X = 1.47253E - 3 Y = 2.53976E - 4
S6 = 1.6 W5 = 2.26023E - 6 W7 = 1.50341E - 3
X = 1.50277E - 3 Y = -4.33987E - 5
```

$$W7 = \sqrt{2}\frac{R}{a}\sqrt{\frac{\tilde{p}_{ff}}{\tilde{p}(w)}}$$

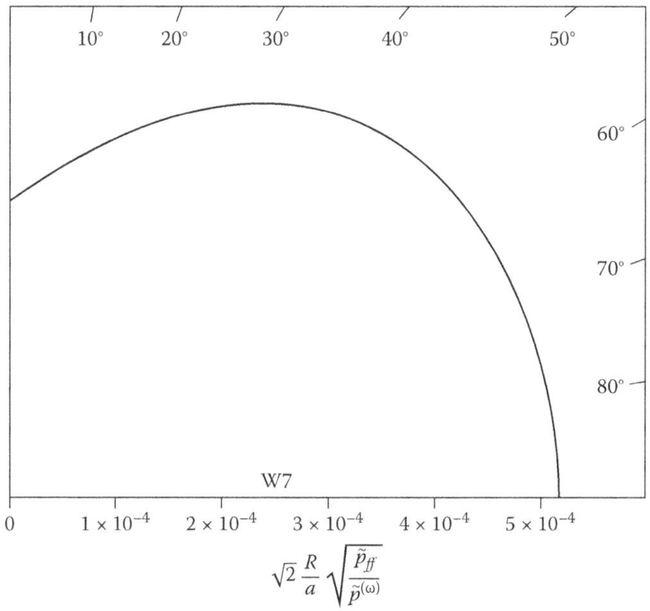

FIGURE 12.2
Directivity pattern $w_a/c_o = 1.0$.

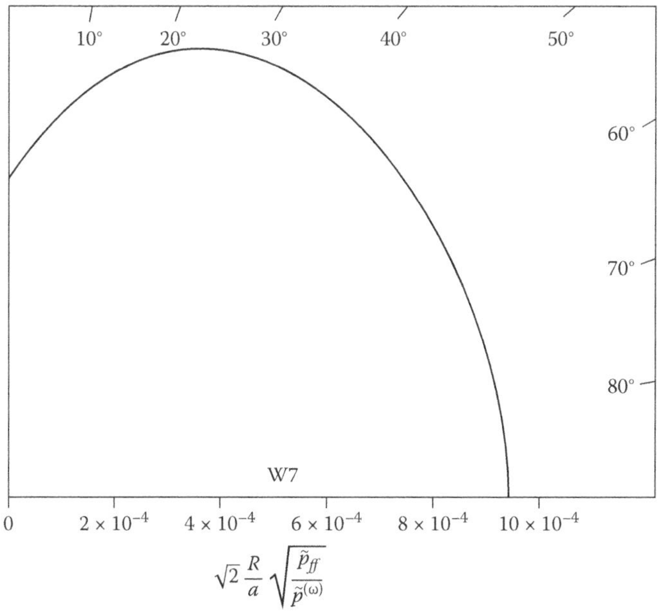

FIGURE 12.3
Directivity pattern $w_a/c_o = 2.0$.

$$W7 = \sqrt{2}\,\frac{R}{a}\,\sqrt{\frac{\tilde{p}_{ff}}{\tilde{p}^{(\omega)}}}$$

W7 is program output

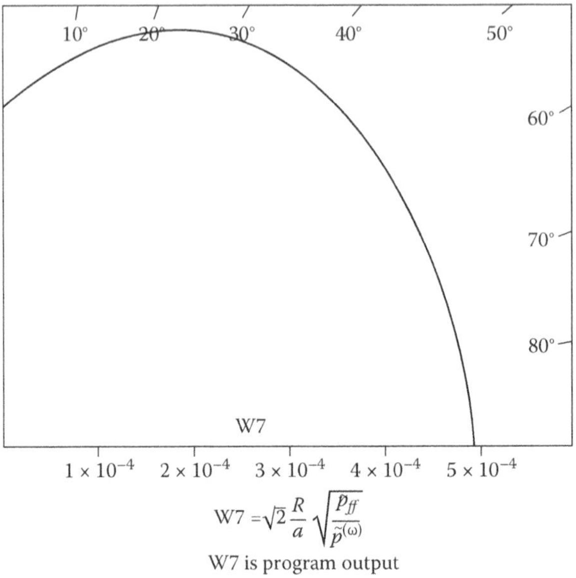

FIGURE 12.4
Directivity pattern $w_a/c_o = 3.0$.

$$\sqrt{2}\,\frac{R}{a}\,\sqrt{\frac{\tilde{p}_{ff}}{\tilde{p}^{(w)}}}$$

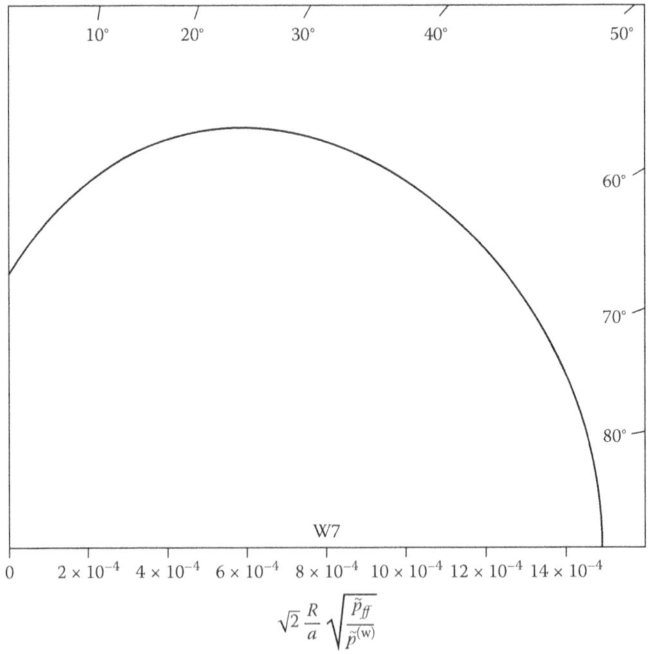

FIGURE 12.5
Directivity pattern $w_a/c_o = 4.0$.

References

1. Arnold, R. N., and Warburton, G. B. 1953. *Proceedings of Institution of Mechanical Engineers* 167:62–64.
2. Smith, P. W., Jr., and Kerwin, E. M., Jr. 1965. Underwater sound radiation from a finite cylinder. Bolt Beranek and Newman, Inc., report no. 1229, contract Nonr–4476(00) (FBM).
3. Greenspon. J. E. 1970. *Random structural acoustics,* vol. I. J G Engineering Research Associates (sponsored by Office of Naval Research).
4. Crandall, S. H., and Yildiz, A. 1962. Random vibration of beams. *Journal of Applied Mechanics* June: 267.
5. Rattayya, J. V., and Junger, M. C. 1964. Flow excitation of cylindrical shells and associated coincidence effects. *Journal of Acoustical Society of America* 36 (5): 878.
6. Strawderman, W. A., and Brand, R. S. 1969. Turbulent-flow-excited vibration of a simply supported, rectangular flat plate. *Journal of Acoustical Society of America* 45 (1): 177.
7. Bozich, D. J. 1964. Spatial correlation in acoustic–structural coupling. *Journal of Acoustical Society of America* 36 (1): 52.
8. Greenspon, J. E. 1967. Far-field sound radiation from randomly vibrating structures. *Journal of Acoustical Society of America* 41 (5): 1201–1207.
9. P. M., and Ingard, K. U. 1961. Linear acoustic theory. In *Encyclopedia of physics,* Vol. XI/1, Acoustics. New York: Springer–Verlag.

13

Applications of Statistical Acoustics to Near Field–Far Field Problems

13.1 The Near Field–Far Field Problem

The equations derived in the previous chapters have been presented from the input–output or transfer function approach. The value of pressure or spectrum in the field was presented in terms of values on a surface. In a theoretical sense, neither the values of pressure on a surface nor field values are known; these relations are merely integral equations that are the counterpart of the differential equations of acoustics. However, if one takes a different view of these relations, it is evident that if the values over a surface are known or can be measured, then the field values can be computed. Conversely, if the field values are known, then the values on the surface can be approximated. With this approach in mind, several applications of the fundamental equations shall be considered.

13.2 Parrent's Solution for the Plane[1,2]

If measurements of pressure alone can be taken over the surface of a plane, then Equation 10.7 can be employed to determine the radiation and the Green's function, which vanishes over the plane, should be used. The geometry for a plane radiating source is shown in Figure 13.1.

The appropriate transfer functions are[3]

$$
\begin{aligned}
\frac{\partial g_1^*(P,S_i,\omega)}{\partial n_i} &= -\frac{2(1-ikr_1)}{4\pi}\cos\theta_1\frac{e^{ikr_1}}{r_1^2} \\
\frac{\partial g_1(Q,S_r,\omega)}{\partial n_r} &= -\frac{2(1+ikr_2)}{4\pi}\cos\theta_2\frac{e^{-ikr_2}}{r_2^2}
\end{aligned}
\tag{13.1}
$$

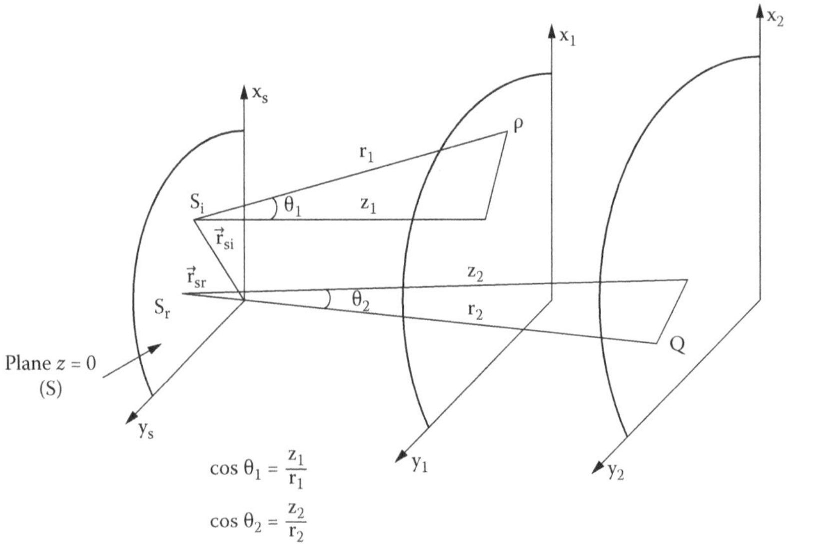

FIGURE 13.1
Parameters for a plane radiating source.

The expression for the cross spectrum of the field pressure then becomes (Equation 10.7):

$$G(P,Q,\omega) = \frac{1}{4\pi^2}\int_S\int_S G(S_i,S_r,\omega)(1-ikr_1)(1+ikr_2)\frac{z_1}{r_1}\frac{z_2}{r_2}\frac{e^{ik(r_1-r_2)}}{r_1^2 r_2^2}dS_i\,dS_r \quad (13.2)$$

Note that r_1 and r_2 are the distances from the surface points to the field points. Beran and Parrent[4] integrate this equation for the extreme cases of coherence and incoherence of the sources on the plane.

13.3 Other Applications of the Parrent Equation

It is very difficult to solve noise problems for surfaces other than the plane since the Green's function cannot be put into such simple form for other practical surfaces. In the case of plane waves, the solution will be given in Section 13.6 for a general surface. Several authors have worked on the solution for the sphere, the cylinder, and the spheroid. Ferris[5] has presented the form of the solution in spheroidal coordinates using the Green's function,[6] which vanishes over the surface. Hudimac and Fitzgerald[7] have suggested the use of the Papas Green's function[8] for the cylinder. Murphy[9] has carried

TABLE 13.1

Formal Solutions Based on Existing Green's Functions

Physical Problem	Ref.	Ref. for Obtaining Green's Function
Plane: pressure[a]	1, 2	2
Plane: acceleration[b]		10
Cylinder: pressure	7	8
Cylinder: acceleration		11
Prolate spheroid: pressure	5	6, 12
Prolate spheroid: acceleration		12

[a] This means that the far field cross spectrum of pressure is computed in terms of cross spectrum of pressure on the surface.

[b] This means that the far field cross spectrum of pressure is computed in terms of cross spectrum of acceleration on the surface.

the work on the sphere and cylinder to a point of obtaining both theoretical and experimental results for single-frequency excitation. Some of the formal solutions based on existing Green's functions are listed in Table 13.1.

13.4 Solution Using a Single Integration with Known Coherence

Another method for computing far field noise from near field measurements, which is slightly different than the Parrent equation, was introduced by Marsh[13] and briefly followed up by Greenspon.[14] This method uses the properties of Equation 10.29. As explained at the end of Section 10.2, the equation has one of the variables as a reference while all operations are done on the other. Thus, if $\bar{P}$ is a readily accessible point in the near field and if Q is in the far field, the second of the equations in Equation 10.29 without the source term becomes

$$\nabla_Q^2 G(\bar{P},Q,\omega) + k^2 G(\bar{P},Q,\omega) = 0 \tag{13.3}$$

where ∇_Q means that the differentiation is taken in the coordinates where Q is the variable point (i.e., Q can take on values for either field or surface point). The single-frequency relation can then be applied; thus,

$$G(\bar{P},Q,\omega) = \frac{1}{4\pi}\int_S \left[G(S,\bar{P},\omega)\frac{\partial g}{\partial n} - g(Q,S,\omega)\frac{\partial G(S,\bar{P},\omega)}{\partial n} \right] dS \tag{13.4}$$

where $G(S,\bar{P},\omega)$ is the cross spectrum between pressure at the reference point $\bar{P}$ and the surface point S on the surface surrounding the sources.

This gives only the cross spectrum between a reference point in the near field $\bar{P}$ and a point in the far field Q. What is desired is an expression for cross spectrum or autospectrum in the far field. The coherence $C_{12}(\omega)$ is defined as

$$C_{12}(\omega) = \frac{|G_{12}(\omega)|^2}{G_{11}(\omega)G_{22}(\omega)} \tag{13.5}$$

where 1 and 2 are any points in the field. $G_{12}(\omega)$ is the cross spectrum between points 1 and 2, respectively. If the field is such that at frequency ω, C_{12} is approximately constant (which would be the case with single-frequency or coherent sources in which $C_{12} = 1$), then

$$G(Q,Q,\omega) = \frac{|G(\bar{P},Q,\omega)|^2}{G(\bar{P},\bar{P},\omega)\ C(\bar{P},Q,\omega)} \tag{13.6}$$

The relations derived in this section will be most useful for fields that are almost coherent.

13.5 Determination of Far Field from Near Field Autospectrum Alone and the Inverse Far Field–Near Field Problem

The formal work of Beran and Parrent[1] and the equations derived in previous chapters of this book have all indicated that the spectrum in the field is determined by the cross spectrum of the pressure (or acceleration) over the surface surrounding the noise sources. In most practical cases, it is very difficult, if not impossible, to obtain the cross spectra for all of the pairs of points over a surface surrounding the noise sources. A number of years ago, an idea that makes it convenient to obtain the spectrum in the field by simply measuring spectrum over the surface surrounding the sources was advanced by Meggs[15] and later, independently, by Butler.[16,17] Actually, Butler[16] used a spherical wave expansion to represent the sound field rather than any formalism associated with the Green's function. Nevertheless, the reason that the method works is because of the separability of the Green's function, as explained next.

The Green's function for a finite structure can be written in the form of

$$g(P,S,\omega) = \sum_j f_j(P,\omega)h_j(S,\omega) \tag{13.7}$$

where $f_j(P,\omega)$ is a function of the coordinates of the field point P, and $h_j(S,\omega)$ is a function of the coordinates of the surface point S. Differentiating Equation 13.7,

$$\frac{\partial g(P,S,\omega)}{\partial n} = \sum_j f_j(P,\omega)\frac{\partial h_j(S,\omega)}{\partial n} \tag{13.8}$$

Substituting into Equation 10.7,

$$G(P,Q,\omega) = \int_{S_i}\int_{S_r} G(S_i,S_r,\omega)\left[\sum_j\sum_k f_j(P,\omega)f_k^*(Q,\omega)\frac{\partial h_j(S_i,\omega)}{\partial n_i}\frac{\partial h_k^*(S_r,\omega)}{\partial n_r}\right]dS_i\, dS_r \tag{13.9}$$

The autospectrum can be written (letting $P = Q$) as

$$G(P,P,\omega) = \sum_j\sum_k [f_j(P,\omega)f_k^*(P,\omega)]\left[\int_{S_i}\int_{S_r} G(S_i,S_r,\omega)\frac{\partial h_j(S_i,\omega)}{\partial n_i}\frac{\partial h_k^*(S_r,\omega)}{\partial n_r}dS_i\, dS_r\right] \tag{13.10}$$

Written more compactly,

$$G(P,P,\omega) = \sum_j\sum_k [F_{jk}(P,\omega)][H_{jk}(\omega)] \tag{13.11}$$

Equation 13.11 can be written as a series of $j \times j$ ($= k \times k$) simultaneous equations for the unknown coefficients $H_{jk}(\omega)$ if we have $j \times j$ ($= j^2$) measurements of $G(P,P,\omega)$. Physically, the coefficients $H_{jk}(\omega)$ are functions of the near field alone (i.e., functions only of S_i, S_r) and $F_{jk}(P,\omega)$ are functions only of the far field coordinates. Thus, if the $H_{jk}(\omega)$ are obtained, the entire field is defined since $F_{jk}(\omega)$ is a known function.

This can be furthered one step. Write $H_{jk}(\omega)$ as a summation as follows:

$$H_{jk}(\omega) = \sum_i\sum_r G(i,r,\omega)\frac{\partial h_j(i,\omega)}{\partial n_i}\frac{\partial h_k^*(r,\omega)}{\partial n_r}\Delta S_i\,\Delta S_r \tag{13.12}$$

where S_i has been replaced by i, S_r by r, and ds_i, ds_r by ΔS_i, ΔS_r. Equation 13.12 can be written as a series of $j \times k = i \times r$ ($= j^2 = r^2$) equations for the unknown cross spectra $G(i, r, \omega)$.

Thus, it is found that, by measurement of $G(P, P, \omega)$ and several transformations, a set of cross spectra that produced this field can be obtained. Butler[17] has demonstrated the soundness of using autospectra (or intensity) to obtain the coefficients in the expansion and then using these coefficients to compute the far field directivity pattern. By using a spherical wave expansion to obtain the field from three point sources with nine hydrophone measurements, he obtains excellent results. (Butler does not use the Green's function expansion in spherical coordinates. However, his expansion is equivalent because the function $f_j(P,\omega)$ in Equation 13.10 is the same as his function and the rest of the expansion is lumped into the unknown coefficient H_{jk}.)

13.6 Simplification of Equations for Plane Waves

Relations 10.7 and 10.8 are completely general. However, they depend upon determining the Green's function g. This is tantamount to solving the whole acoustics problem. This is true since we must determine a function that satisfies the Helmholtz equation with a δ function on the right side, subject to boundary conditions of the function itself vanishing on the boundary or its normal derivative vanishing on the boundary. Suppose that it is assumed, as was done by Horton and Innis,[18] that the pressure can be approximated by plane waves in the immediate vicinity of the surface. The pressure will then be of the following form[19]:

$$p(P,t) = p(\rho,t) \tag{13.13}$$

A plane wave is one that is constant over each of the planes ρ = constant (see Figure 13.2). For this case, the wave equation becomes

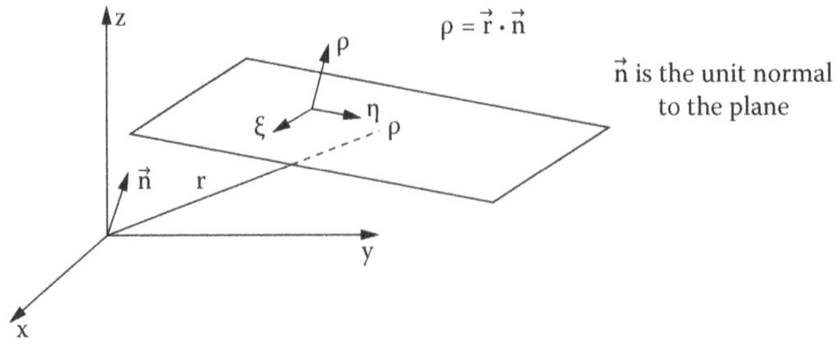

FIGURE 13.2
Plane wave coordinates.

$$\frac{\partial^2 p}{\partial \rho^2} = \frac{1}{C^2}\frac{\partial^2 p}{\partial t^2} \tag{13.14}$$

This is true because ∇^2 is transformed from the x, y, z system to the ξ, η, ρ system and, in this latter system, the wave is not a function of ξ, η. The general solution of Equation 13.14 is

$$p = f_1(\rho - Ct) + f_2(\rho + Ct) \tag{13.15}$$

Thus, the waves are propagated in the ρ direction with velocity C. Furthermore, if the wave is harmonic, then

$$p(\rho, t) = p(\rho)e^{i\omega t} \tag{13.16}$$

so that

$$\frac{d^2 p(\rho)}{d\rho^2} + k^2 p(\rho) = 0 \qquad k^2 = \omega^2/C^2 \tag{13.17}$$

Therefore, the pressure is proportional to $e^{ik\rho}$ or $e^{-ik\rho}$ and

$$\frac{\partial p}{\partial n} = \frac{\partial p}{\partial \rho} = ikp \tag{13.18}$$

Choose

$$g(P, S, \omega) = \frac{1}{4\pi}e^{ikR} \tag{13.19}$$

where R is the distance from S to P; that is, if x, y, and z are the rectangular coordinates of P and x_s, y_s, and z_s are the rectangular coordinates of S, then

$$R = \sqrt{(x - x_s)^2 + (y - y_s)^2 + (z - z_s)^2} \tag{13.20}$$

Substituting Equation 11.18 and Equation 11.19 into Equation 9.17,

$$\begin{aligned} p(P,\omega) &= \frac{1}{4\pi}\int_{S_0}\left[p(S,\omega)\frac{\partial}{\partial n}\left(\frac{e^{ikR}}{R}\right) - \frac{e^{ikR}}{R}ikp(S,\omega)\right]dS \\ &= \frac{1}{4\pi}\int_{S_0}[p(S,\omega)]\left[\frac{ik}{R}e^{ikR}\left(\frac{\partial R}{\partial n} - 1\right) - \frac{e^{ikR}}{R^2}\frac{\partial R}{\partial n}\right]dS \end{aligned} \tag{13.21}$$

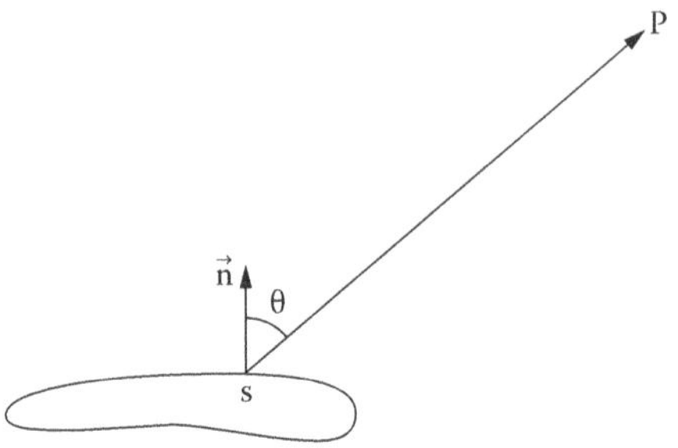

FIGURE 13.3
Angle to field point.

Horton and Innis[18] have further simplified this relation for P in the Fraunhofer zone (the far field). For this case, the term containing R^2 in the denominator can be neglected and

$$\frac{\partial R}{\partial n} = -\cos\theta \tag{13.22}$$

where θ is the angle between the normal vector $\vec{n}$ and the vector pointing from S to P (see Figure 13.3).

Under these circumstances, Equation 13.21 reduces to

$$p(P,\omega) = -\frac{1}{4\pi}\int_{S_0} [p(S,\omega)]\left[\frac{ik}{R}e^{ikR}(1+\cos\theta)\right]dS \tag{13.23}$$

Even for cylindrical and spherical waves, based on analysis of Müller,[2] Horton and Innis[19] have pointed out that (for spherical waves)

$$\frac{\partial p}{\partial n} = ikp + \in(k_r) \tag{13.24}$$

where $\in$ is of order $1/(kr)^2$.

Thus, if kr is sufficiently large,

$$\frac{\partial p}{\partial n} \approx ikp.$$

Comparing Equation 13.23 with Equation 10.2, it is seen that for plane waves and for spherical or cylindrical waves in which $kr \gg 1$, if P is in the far field, then

$$H_p(P,S,\omega) \approx \frac{1}{4\pi}\left[\frac{ik}{R}e^{ikR}(1+\cos\theta)\right] \tag{13.25}$$

and the cross spectrum of far field pressure in terms of surface pressure as given by Equation 10.4 follows immediately as

$$G(P,Q,\omega) = \frac{1}{(4\pi)^2}\int_{S_i}\int_{S_r}\left[-\frac{ik}{R_P}e^{-ikR_P}(1+\cos\theta_P)\right]\left[\frac{ik}{R_Q}e^{ikR_Q}(1+\cos\theta_Q)\right]G(S_i,S_r,\omega)dS_i\ dS_r \tag{13.26}$$

Similarly, for the plane wave (spherical or cylindrical $kr \gg 1$), we could also write

$$p = \frac{1}{ik}\frac{\partial\rho}{\partial n} \tag{13.27}$$

and

$$p(P,\omega) = \frac{1}{4\pi}\int_S\left[\frac{1}{ik}\frac{\partial p}{\partial n}\frac{\partial}{\partial n}\left(\frac{e^{ikR}}{R}\right) - \frac{e^{ikR}}{R}\frac{\partial p}{\partial n}\right]dS \tag{13.28}$$

$$= \frac{1}{4\pi}\int_S\frac{\partial p}{\partial n}\left[\frac{e^{ikR}}{R}\left(\frac{1}{ik}\frac{\partial R}{\partial n} - 1\right) - \frac{e^{ikR}}{ikR^2}\frac{\partial R}{\partial n}\right]dS \tag{13.29}$$

For the Fraunhofer zone,

$$p(P,\omega) = -\frac{1}{4\pi}\int_S\frac{\partial p}{\partial n}\left[\frac{e^{ikR}}{R}(1+\cos\theta)\right]dS \tag{13.30}$$

Using Equation 10.3 and $\dfrac{\partial p}{\partial n} = -\rho a_n$,

$$H_a(P,S,\omega) = \rho\frac{e^{ikR}}{R}(1+\cos\theta) \tag{13.31}$$

and the cross spectrum in terms of surface motion, as given by Equation 10.5, becomes

$$G(P,Q,\omega) = \frac{1}{(4\pi)^2}\int_{S_i}\int_{S_r}\rho^2\left[\frac{e^{-ikR_P}}{R_P}(1+\cos\theta_P)\right]\left[\frac{e^{ikR_Q}}{R_Q}(1+\cos\theta_Q)\right]A(S_i,S_r,\omega)dS_i\ dS_r \tag{13.32}$$

13.7 Methods for Computing Far Field from Near Field Acceleration

13.7.1 The Direct Method

The pressure p generated by any noise source satisfies the wave equation as given by

$$\nabla^2 p = \frac{1}{c^2}\frac{\partial^2 p}{\partial t^2}, \quad p = p(x,y,z,t) \tag{13.33}$$

For a general noise, p can be represented by a Fourier transform as follows:

$$p(x,y,z,t) = \frac{1}{\sqrt{2\pi}}\int_{+\infty}^{-\infty} U(x,y,z,\omega)e^{-i\omega t}d\omega \tag{13.34}$$

where U is the Fourier transform of p, t is the time, and ω is the frequency (which is considered a continuous variable for a transient or broadband process). Thus, by the Fourier inversion theorem,

$$U(x,y,z,\omega) = \frac{1}{\sqrt{2\pi}}\int_{+\infty}^{-\infty} p(x,y,z,\omega)e^{i\omega t}dt \tag{13.35}$$

Substituting Equation 13.34 into Equation 13.33, we find that U satisfies the ordinary Helmholtz equation

$$\nabla^2 U + k^2 U = 0, \quad k^2 = \frac{\omega^2}{c^2} \tag{13.36}$$

The Kirchoff theorem states that if all of the sources of sound are interior to a closed surface, S, then the Fourier transform, $U(P,\omega)$, in the field at point P and frequency ω can be written in terms of the Fourier transform of the surface acceleration (normal to the surface) over the closed surface as follows (this assumes that the plane wave approximation can be made; see pages 25–27 of Greenspon[21]):

$$U(P,\omega) = \frac{\rho_0}{2}\int_S \left[\frac{e^{-ikR_p}}{R_p}\right](1+\cos\theta_p)A(S,\omega)dS \tag{13.37}$$

where $A(S,\omega)$ is the Fourier transform of the surface acceleration and ρ_0 is the density of the medium.

The spectrum of the far field pressure, $S_p(P,\omega)$, at point P and frequency ω is then (U_T denotes the truncated Fourier transform):

$$S_p(P,\omega) = \lim_{T\to\infty} \frac{|U_T(P,\omega)|^2}{T} \tag{13.38}$$

For single frequency, Equation 13.37 holds, except that $U(P,\omega)$ now becomes the far field pressure and $A(S,\omega)$ is the acceleration on the surface in magnitude and phase. For a transient or random process (narrow or broad band), Equation 13.37 holds as it was written previously.

Simple numerical procedures for calculating the surface integral in Equation 13.37 if the closed surface is cylindrical can be found in Baker.[24]

13.7.2 Point Force Method

This method applies to physical phenomena such as transients resulting from localized excitation at a single point. It has been shown[21] that the spectrum of the radiated field pressure can be represented as follows:

$$S_p(\vec{r}_1,\omega) = \int_{\vec{r}_0}\int_{\vec{r}_0'} d\vec{r}_0\, d\vec{r}_0'\, G^*(\vec{r}_1,\vec{r}_0,\omega)\; G(\vec{r}_1,\vec{r}_0,\omega)\; S_L(\vec{r}_0,\vec{r}_0',\omega) \tag{13.39}$$

where

$S_p(r_1,\omega)$ = spectrum of the radiated field pressure at field point r_1 at frequency ω

$G^*(r_1,r_0,\omega)$ = complex conjugate of pressure at field point r_1 due to unit sinusoidal force at frequency ω at surface point r_0 (point of excitation)

$G(r_1,r_0',\omega)$ = pressure at field point r_1 due to unit sinusoidal force at frequency ω at surface point r_0'

$S_L(r_0,r_1',\omega)$ = cross spectrum of the force at frequency ω between surface points r_0 and r_0'

For a point load at r_p,

$$S_L(\vec{r}_0,\vec{r}_0',\omega) = S_F(\omega)\delta(\vec{r}_0-\vec{r}_p)\delta(\vec{r}_0'-\vec{r}_p) \tag{13.40}$$

where δ represents the delta function and $S_F(\omega)$ is the spectrum of the force at exciting point p. Substituting Equation 13.40 into Equation 13.39, we obtain

$$S_P(\vec{r}_1,\omega) = S_F(\omega)\left|G(\vec{r}_1,\vec{r}_p,\omega)\right|^2 \tag{13.41}$$

where $G(r_1,r_p,\omega)$ is the pressure at field point r_1 due to unit sinusoidal force at frequency ω at surface point r_p.

Thus, if $G(r_1,r_p,\omega)$ can be determined for general cases, then all that has to be done is to place a force transducer on the shell at each point force to measure $S_F(\omega)$ and then apply Equation 13.41 to determine the complete field pressure spectrum.

The spectrum of the force can be obtained from the spectrum of acceleration. The spectrum of the force at the exciting point, $S_F(\omega)$, can be written in terms of the spectrum of the acceleration at that point, $S_A(\omega)$, by the following relation[8]:

$$S_A(\omega) = \left|H_{AF}(\omega)\right|^2 S_F(\omega) \tag{13.42}$$

where H_{AF} is the velocity at the exciting point due to unit sinusoidal force at the exciting point.

13.8 Application to Prediction of Radiation

13.8.1 Direct Method Calculations

13.8.1.1 Far Field Fourier Transform[24]

The coordinate system is as shown in Figure 13.4. Equation 13.37 can be written for the far field Fourier transform as follows:

$$U(P,\omega) = \frac{\rho_0}{4\pi}\frac{e^{ikR}}{R}\int_0^{2\pi}[1+\cos\varphi\cos(\psi-\alpha)]e^{-ika\cos\varphi\cos(\psi-\alpha)}\int_{+l/2}^{-l/2} p(z,\alpha)e^{-ikz\sin\varphi}a\,dz\,d\alpha \tag{13.43}$$

The integrals are evaluated by the rectangular method:

$$U(R,\varphi,\psi,\omega) = \left[\frac{\rho_0}{4\pi}\frac{e^{ikR}}{R}\frac{\Delta l a\Delta\alpha}{2}\sum_{m=1}^{M}[1+\cos\varphi\,\cos(\psi-\alpha_m)]e^{-ika\cos\varphi\,\cos(\psi-\alpha_m)}\right]\left[\sum_{n=1}^{N}A_n p(z_n,\alpha_m)e^{-ikz_n\sin\varphi}\right] \tag{13.44}$$

where $a\Delta\alpha$ is the angular increment, Δl is the length increment, and $M = [2\pi/\Delta\alpha]$.

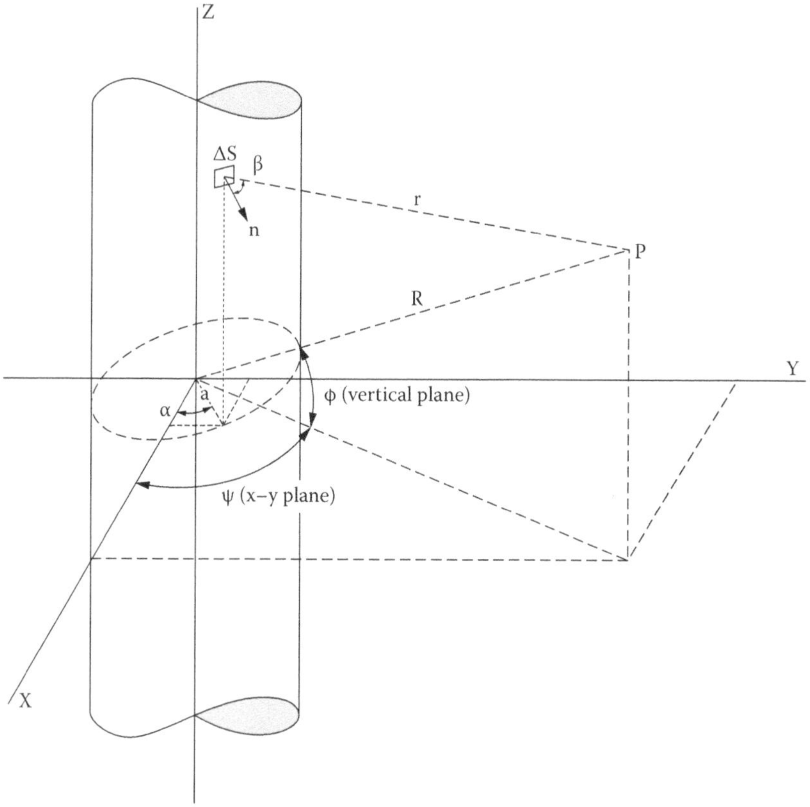

FIGURE 13.4
Coordinate system used in the direct method.

The acceleration at the unknown points is evaluated by using the solution for the point load on a cylinder[22] at high frequency. The normal displacement, w, on the surface of the cylinder is given as follows:

$$w(a,z,\theta)=\frac{iF}{\omega Z_p}\sum_{m=0}^{+\infty}\left(\frac{2}{\pi k_f R_m}\right)^{\frac{1}{2}} e^{i\left(k_f R_m-\frac{\pi}{4}\right)}+\left(\frac{2}{\pi k_f R'_m}\right)^{\frac{1}{2}} e^{i\left(k_f R'_m-\frac{\pi}{4}\right)} \tag{13.45}$$

where
$R_m = (z^2 + a^2(\theta + 2m\pi)^2)^{1/2}$
$R'_m = (z^2 + a^2(2\pi - \theta + 2m\pi)^2)^{1/2}$
$k_f = (\omega/c_p h)(12)^{-1/4}$
$Z_p = 4mc_p h/(3)^{1/2}$
a = radius of the cylinder
z = longitudinal coordinate of the point at which the displacement is measured

θ = angular coordinate of the point at which the displacement is measured
ω = circular frequency ($2\pi f$)
f = frequency in hertz
$c_p = (E/\rho(1 - \upsilon^2))^{1/2}$ (the plate velocity)
E = modulus of elasticity of the cylinder material
ρ = mass density of the cylinder material
ν = Poisson's ratio of the cylinder material
m = mass per unit area of the cylinder

13.9 Inverse Method Calculation: Development of the Field in Terms of Point Sources

13.9.1 Governing Equations

Consider a surface surrounding M simple sources that produce a total acoustic pressure, p, at time t and location r_j. Thus,

$$p(\vec{r}_j, t) = \sum_{m=1}^{M} \frac{f_m(t - \frac{r_{jm}}{C})}{r_{jm}} \tag{13.46}$$

where C is the speed of sound in the medium and r_{jm} is the distance between the nth source and the jth field point, as shown in Figure 13.5.

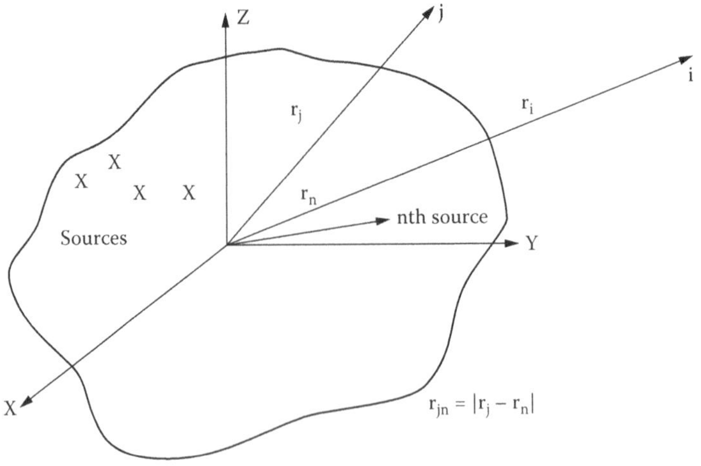

FIGURE 13.5
Sources and field points.

The cross correlation $\Gamma(r_i, r_j, \tau)$ at field points i and j is defined as

$$\Gamma(\vec{r}_i, \vec{r}_j, \tau) = \lim_{T\to\infty} \frac{1}{2T} \int_{+T}^{-T} p(\vec{r}_i, t) p(\vec{r}_j, t+\tau) dt \tag{13.47}$$

The cross spectrum, $G(r_i, r_j, \omega)$, is the Fourier transform of the cross correlation; thus,

$$G(\vec{r}_i, \vec{r}_j, \omega) = \frac{1}{\sqrt{2\pi}} \int_{-\infty}^{+\infty} \Gamma(\vec{r}_i, \vec{r}_j, \tau) e^{-i\omega\tau} d\tau \tag{13.48}$$

Since the Fourier transform of a convolution is equal to the product of the transforms of the separate functions, we have

$$G(\vec{r}_i, \vec{r}_j, \omega) = \lim_{T\to\infty} \frac{1}{2T} G_T(\vec{r}_j, \omega)\ G_T^*(\vec{r}_i, \omega) \tag{13.49}$$

where the subscript T denotes that the transform is the truncated transform.

Thus,

$$G_T(\vec{r}_j, \omega) = \frac{1}{\sqrt{2\pi}} \int_{-\infty}^{+\infty} p_T(\vec{r}_j, t) e^{-i\omega t} dt \tag{13.50}$$

Now, substituting Equation 13.52 into Equation 13.50, we obtain

$$G_T(\vec{r}_j, \omega) = \sum_{m=1}^{M} F_{mT}(\omega) \frac{e^{(i\frac{\omega}{c} r_{jm})}}{r_{jm}} \tag{13.51}$$

$$F_{mT} = \frac{1}{\sqrt{2\pi}} \int_{-\infty}^{+\infty} f_{mT}(t) e^{-i\omega t} dt \tag{13.52}$$

$$G_T(\vec{r}_i, \omega) = \sum_{m=1}^{M} F_{mT}(\omega) \frac{e^{(i\frac{\omega}{c} r_{im})}}{r_{im}} \tag{13.53}$$

With integration, we have

$$G_T(\vec{r}_i,\vec{r}_j,\omega) = \sum_{m=1}^{M}\sum_{n=1}^{M} g_{nm}(\omega)\frac{e^{[i\frac{\omega}{c}(r_{jn}-r_{im})]}}{r_{jn}r_{im}} \tag{13.54}$$

and the autospectrum at the point j will be

$$G_T(\vec{r}_j,\vec{r}_j,\omega) = \sum_{m=1}^{M}\sum_{n=1}^{M} g_{nm}(\omega)\frac{e^{[i\frac{\omega}{c}(r_{jn}-r_{jm})]}}{r_{jn}r_{jm}} \tag{13.55}$$

where

$$g_{nm}(\omega) = \lim_{T\to\infty}\frac{1}{2T}F_{nT}(\omega)F^{*}_{mT}(\omega) \tag{13.56}$$

The $g_{mn}(\omega)$ are the cross spectra among the sources. If we have M^2 measurements of $G(r_j, r_j, \omega)$, then Equation 13.55 contains a series of M^2 complex algebraic equations for the unknown cross spectra $g_{mn}(\omega)$. Once these source cross spectra, $g_{mn}(\omega)$, are obtained, then the field can be computed at any other field point by substituting the appropriate r_{jn} and r_{jm} into Equation 13.55.

13.10 Point Force Calculations

Section 13.7 showed that the spectrum of the far field pressure is dependent upon the pressure at the field point due to a unit sinusoidal input at the force point. This is called Green's function and was derived by Junger and Feit in their treatise.[23] This result is shown in the following equation:

$$G(\vec{r}_1,\vec{r}_p,\omega) = -\frac{ik}{2\pi R}e^{ik(R-a\sin\theta\cos\phi)} \frac{\sin\theta\cos\phi}{1-[i\omega\rho_s h/(\rho c)]\;\cos\phi\sin\theta\{1-(\omega^2/\omega_c^2)\;[\cos^2\theta+\sin^2\theta\sin^2\phi]^2\}} \tag{13.57}$$

where

$G(r_1,r_p,\omega)$ = far field pressure at point r_1 due to a unit sinusoidal load at surface point r_p

ω = circular frequency ($2\pi f$, f = frequency in hertz)

R,θ,φ = spherical coordinates of the far field point
ρ_s = mass density of shell material
ρ = mass density of water
c = sound velocity in water
ω_c = coincidence frequency = $((12)^{1/2} c^2/hc_p)$
h = thickness of shell wall (an equivalent thickness that takes account of stiffeners)
c_p = plate velocity = $(E/\rho_s(1-\nu^2))^{1/2}$
E = modulus of elasticity of shell material
ν = Poisson ratio for shell material

Equation 13.42 states that

$$S_A(\omega) = |H_{AF}(\omega)|^2 S_F(\omega)$$

where
$S_F(\omega)$ = spectrum of the applied force
$S_A(\omega)$ = spectrum of the acceleration
$H_{AF}(\omega)$ = acceleration at the exciting point due to unit sinusoidal force at the exciting point

Equation 13.41 states that

$$S_p(\vec{r}_1,\omega) = S_F(\omega)|G(\vec{r}_1,\vec{r}_p,\omega)|^2 \quad (13.58)$$

where $S_P(r_1, \omega)$ = spectrum at the far field pressure at point r_1 and the other variables are as described previously.

References

1. Beran, M. J., and Parrent, G. B. Jr., 1964. *Theory of partial coherence.* Englewood Cliffs, NJ: Prentice Hall, Inc.
2. Parrent, G. B., Jr. 1959. *Journal of Optical Society of America* 49: 787.
3. Beran, M. J., and Parrent, G. B. Jr., 1964. *Theory of partial coherence,* 42. Englewood Cliffs, NJ: Prentice Hall, Inc.
4. Beran, M. J., and Parrent, G. B. Jr., 1964. *Theory of partial coherence.* Englewood Cliffs, NJ: Prentice Hall, Inc.
5. Ferris, H. G. 1964. Farfield radiation pattern of a noise source from nearfield measurements. *Journal of Acoustical Society of America* 36: 1597–1598.
6. Morse, P. M., and Feshbach, H. 1953. *Methods of theoretical physics,* vol. I, 892. New York: McGraw–Hill Book Co.
7. Hudimac, A. A., and Fitzgerald, E. M. 1964. *Journal of Underwater Acoustics* 14:645–661.

8. Papas, C. H. 1950. *Journal of Mathematical Physics* 28: 227.
9. Murphy, D. A. 1969. Far field prediction of submarine radiated noise from near field measurements—Vol. II. Hughes Aircraft Co., contract N00014–69–C–0034, 20.
10. Sneddon, I. 1957. *Elements of partial differential equations,* 244–245. New York: McGraw–Hill Book Co., Inc.
11. Junger, M. C. 1954. Theory and design of an end–fire directive sound source. Acoustics Research Laboratory, Harvard University, Cambridge, MA, contract N50RI–76 tech. memo. no. 34, p. 60.
12. Flammer, C. 1957. *Spheroidal wave functions,* 47. Stanford, CA: Stanford University Press.
13. Marsh, H. W. 1953. Near field–far field noise relations. Suppl. of letter to M. Lasky (Off. Naval Res.) unpublished.
14. Greenspon, J. E. 1964. Farfield-noise prediction from nearfield-spectrum measurements. *Journal of Acoustical Society of America* 36: 604.
15. Meggs, W. J. 1965. U.S.N. Marine Engineering Lab., tech. memo. 422/65.
16. Butler, J. L. 1968. A series expansion method for the prediction of the far field from near field measurements. Parke Math. Labs., contract N00014–67–C–0424, scientific rep. no. 1.
17. Butler, J. L. 1969. A method for the prediction of the far field from near field measurements alone. Parke Math. Labs., contract N00014–67–C–0424, scientific report no. 3.
18. Horton, C. W., and Innis, G. S., Jr. 1961. The computation of far-field radiation patterns from measurements made near the source. *Journal of Acoustical Society of America* 33: 877.
19. Born, M., and Wolf, E. 1959. *Principles of optics,* 14. New York: The MacMillan Co.
20. Muller, C. 1957. *Grundprobleme der Mathematischen Theorie Elektromagnetischen Schwingungen,* 83. Dordrecht, the Netherlands: Springer–Verlag.
21. Greenspon, J.E. 1970. *Random structural acoustics,* vol. 1. J G Engineering Research Associates (work performed under contract Nonr–5040 (00)).
22. Feit, D. 1971. High frequency response of a point excited cylindrical shell. *Journal of Acoustical Society of America* 49 (5, Part 2): 1494–1504.
23. Junger, M. C., and Feit, D. 1986. *Sound, structures and their interaction.* Cambridge, MA: MIT Press.
24. Baker, D. D. 1962. Determination of far field characteristics of large underwater sound transducers from near field measurements. *Journal of Acoustical Society of America* 34 (11): 1737–1744.
25. Vogel, W., and Feit, D. 1980. Response of a point excited infinitely long cylindrical shell immersed in an acoustic medium. David Taylor NSRDC, DTNSRDC, 80061.

14

Scale Models of Random Loading and Response

14.1 Approximate Formulation of the Modeling Laws

The theory presented here is an application and extension of the basic theory of modeling presented by Weber.[2] Let the quantities associated with the full scale be primed and those associated with the model unprimed. Thus, if ℓ' is the characteristic length in the full scale, t' the full-scale time, and k' the force in the full scale,

$$\ell' = \ell\lambda, \qquad t' = t\tau, \qquad k' = k\chi \tag{14.1}$$

where
λ is the linear dimension scale factor
τ is the time scale factor
χ is the force scale factor

The phenomenon that is to be scaled is random loading and response of an elastic structure. The only assumption that will be made is that the model is geometrically similar to the prototype and is subjected to a geometrically similar load distribution.

The basic equations governing the elastic motions of the structure are the three-dimensional equations of elasticity.[3] They are neglecting body forces.

$$(\bar{\lambda} + \bar{\mu})\left(\frac{\partial \Delta}{\partial x}, \frac{\partial \Delta}{\partial y}, \frac{\partial \Delta}{\partial z}\right) + \bar{\mu}\nabla^2(u, v, w) = \rho_s\left(\frac{\partial^2 u}{\partial t^2}, \frac{\partial^2 v}{\partial t^2}, \frac{\partial^2 w}{\partial t^2}\right) \tag{14.2}$$

where
$\bar{\lambda}$ and $\bar{\mu}$ are the Lamé constants
Δ is the cubical dilatation
ρ_s is the mass density of the solid material

The displacements are u,v,w. The cubical dilatation and the Laplacian are

$$\Delta = \frac{\partial u}{\partial x} + \frac{\partial v}{\partial y} + \frac{\partial w}{\partial z}, \quad \nabla^2(u,v,w) = \frac{\partial^2(u,v,w)}{\partial x^2} + \frac{\partial^2(u,v,w)}{\partial y^2} + \frac{\partial^2(u,v,w)}{\partial z^2} \tag{14.3}$$

The Lamé constants $\bar{\lambda}$ and $\bar{\mu}$ can be written in terms of the modulus of elasticity E and Poisson's ratio ν as follows:

$$\bar{\lambda} = \frac{E\nu}{(1+\nu)(1-2\nu)}, \quad \bar{\mu} = \frac{E}{2(1+\nu)} \tag{14.4}$$

Therefore, the χ component of the equations of motion becomes

$$\frac{E}{2(1+\nu)(1-2\nu)}\left(\frac{\partial^2 u}{\partial x^2} + \frac{\partial^2 v}{\partial x \partial y} + \frac{\partial^2 w}{\partial x \partial z}\right) + \frac{E}{2(1+\nu)}\left(\frac{\partial^2 u}{\partial x^2} + \frac{\partial^2 u}{\partial y^2} + \frac{\partial^2 u}{\partial z^2}\right) = \rho_s \frac{\partial^2 u}{\partial t^2} \tag{14.5}$$

and there are two other equations for the y and z components.

The scaling laws can be derived merely by working with one of these equations. In accordance with our notation, Equation 14.5 is for the model since it contains unprimed quantities. The equivalent equation for the prototype, assuming a different material with approximately the same Poisson's ratio (usually the Poisson's ratio is of secondary importance; moreover, the Poisson's ratio for the model probably will not vary greatly from the full scale material) but difference E and ρ, is

$$\begin{aligned} &\frac{E'}{2(1+\nu)(1-2\nu)}\left(\frac{\partial^2 u'}{\partial x'^2} + \frac{\partial^2 v'}{\partial x' \partial y'} + \frac{\partial^2 w'}{\partial x' \partial z'}\right) \\ &+ \frac{E'}{2(1+\nu)}\left(\frac{\partial^2 u'}{\partial x'^2} + \frac{\partial^2 u'}{\partial y'^2} + \frac{\partial^2 u'}{\partial t'^2}\right) = \rho_s' \frac{\partial^2 u'}{\partial t'^2} \end{aligned} \tag{14.6}$$

Let the scale factors for E and ρ_s be

$$E'/E = \sigma_E, \quad \rho_s'/\rho_s = \sigma_s \tag{14.7}$$

By the choice of model material and linear scale, this will fix σ_E, σ_{ρ_s}, and λ. The rest of the scale factors will then be derived from these basic quantities. Thus, the time scale factor τ will be determined as follows:

$$E'/E = \sigma_E = \frac{k'/\ell'^2}{k/\ell^2} = \frac{k\chi/\ell^2\lambda^2}{k^2/\ell^2} = \chi/\lambda^2$$
$$\rho_s'/\rho_s = \sigma_{\rho_s} = \frac{k't'^2/\ell'^4}{kt^2/\ell^4} = \frac{k\chi t^2\tau^2/\ell^4\lambda^4}{kt^2/\ell^4} = \chi\tau^2/\lambda^4 \tag{14.8}$$

Thus,

$$\chi = \sigma_E\lambda^2$$

and

$$\tau = \sqrt{\sigma_{\rho_s}\lambda^4/\chi} = \lambda\sqrt{\sigma_{\rho_s}/\sigma_E}$$

Frequency is therefore modeled as $\frac{1}{\tau} = \frac{1}{\lambda}\sqrt{\sigma_E/\sigma_{\rho_s}}$.

Equation 14.6 for the prototype can then be written as

$$\frac{\sigma_E E}{2(1+\nu)}\frac{1}{\lambda}\left[\frac{\partial^2 u}{\partial x^2} + \frac{\partial^2 u}{\partial y^2} + \frac{\partial^2 u}{\partial z^2} + \frac{1}{1-2\nu}\left(\frac{\partial^2 u}{\partial x^2} + \frac{\partial^2 v}{\partial x\partial y} + \frac{\partial^2 w}{\partial x\partial z}\right)\right] = \sigma_{\rho_s}\rho_s\frac{\lambda}{\tau^2}\frac{\partial^2 u}{\partial t^2} \tag{14.9}$$

In order that the same equation will hold for both the model and prototype, the following relation must hold:

$$\sigma_E/\lambda = \sigma_{\rho_s}\lambda/\tau^2$$

or

$$\frac{E'/E}{\rho_s'/\rho_s}\,\frac{t'^2/t^2}{\ell'^2/\ell^2} = 1 \tag{14.10}$$

or

$$\frac{E'}{f'^2\rho_s'\ell'^2} = \frac{E}{f^2\rho_s\ell^2}$$

or, finally,

$$f'\ell'/\sqrt{E'/\rho_s'} = f\ell/\sqrt{E/\rho_s}$$

(f = frequency).

The physical number $f\ell/\sqrt{E/\rho_s}$ is a type of Strouhal number. If the same material were used in the model and prototype, then Junger's relations[4] would hold—that is,

$$f\ell = f'\ell' \tag{14.11}$$

For Junger's case, frequency is scaled inversely to length. Equations 14.8 and 14.10 are the basic relations for modeling with different materials. Variations of this scaling law are contained in references 5–9. The other quantities involved in the problem can be derived from these relations. Thus,

$$\begin{aligned}
&\text{Pressure:} \quad p'/p = \chi/\lambda^2 && p' = \sigma_{Ep} \\
&\text{Stress:} \quad \sigma_x'/\sigma_x = \chi/\lambda^2 && \sigma_x' = \sigma_E\sigma_x \\
&\text{Velocity:} \quad v'/v = \frac{\ell'/t'}{\ell/t} = \frac{\lambda\ell/tz}{\ell/t} = \frac{\lambda}{\tau} = \sqrt{\sigma_E/\sigma_{\rho_s}}
\end{aligned} \tag{14.12}$$

(for the same material the velocities scale directly)

Acceleration

The statistical quantities will then be scaled as follows:

Mean square pressure $\dfrac{\langle p'^2\rangle}{\langle p^2\rangle} = \sigma_E^2$

Correlation function–that is, $\dfrac{C'}{C} = \dfrac{\text{Correlation full scale}}{\text{Correlation model}}$:

$$\begin{aligned}
&\text{(a) pressure: } \langle p_1'p_2'\rangle/\langle p_1p_2\rangle = \sigma_E^2 \\
&\text{(b) Displacement: } \langle w_1'w_2'\rangle/\langle w_1w_2\rangle = \lambda^2 \\
&\text{(c) Velocity: } \langle v_1'v_2'\rangle/\langle v_1v_2\rangle = \sigma_E/\sigma_{\rho_s} \\
&\text{(d) Acceleration: } \langle a_1'a_2'\rangle/\langle a_1a_2\rangle = \frac{1}{\lambda^2}\left(\frac{\sigma_E}{\sigma_{\rho_s}}\right)^2
\end{aligned} \tag{14.13}$$

Power spectral density—that is, $\dfrac{(P.S.)'}{(P.S.)}$:

(a) Pressure: $\sigma_E^2\tau = \sigma_E^2\lambda\sqrt{\sigma_{\rho_s}/\sigma_E} = \lambda\sqrt{\sigma_{\rho_s}\sigma_E^3}$

(b) Displacement: $\lambda^2\tau = \lambda^2\lambda\sqrt{\sigma_{\rho_s}/\sigma_E} = \lambda^3\sqrt{\sigma_{\rho_s}/\sigma_E}$

(c) Velocity: $\sigma_E/\sigma_{p_s}\tau = \sigma_E/\sigma_{p_s}\lambda\sqrt{\sigma_{p_s}/\sigma_E} = \lambda\sqrt{\sigma_E/\sigma_{p_s}}$

(d) Acceleration: $\frac{1}{\lambda^2}\left(\frac{\sigma_E}{\sigma_{p_s}}\right)^2\tau = \frac{1}{\lambda^2}\left(\frac{\sigma_E}{\sigma_{p_s}}\right)^2\lambda\sqrt{\frac{\sigma_{p_s}}{\sigma_E}} = \frac{1}{\lambda}\sqrt{\frac{\sigma_E^3}{\sigma_{p_s}^3}}$

14.2 Damping

14.2.1 Joint or Interface Damping

One of the big problems in using scale models is damping. Here we will analyze how Coulomb damping and material damping enter into the problem. Coulomb damping will arise as external damping through rubbing between attached pieces. The basic relation for Coulomb damping states is that the frictional shearing stress is proportional to the normal stress—that is,

$$(S.S.) = \bar{\bar{\mu}}\ (N.S.) \tag{14.14}$$

where $\bar{\bar{\mu}}$ is the coefficient of sliding friction between the two surfaces. For the prototype, this relation will be

$$(S.S.)' = \bar{\bar{\mu}}'(N.S.)' \tag{14.15}$$

The previous scaling laws derived in Section 14.1 stated that the stress must scale by a factor σ_E; thus,

$$(S.S.)' = \sigma_E(S.S.) \tag{14.16}$$

Thus, the frictional shearing stress in the model must be adjusted to $1/\sigma_E$ of the frictional shearing stress in the prototype. This can be done by adjusting the coefficient of friction and the normal bearing stress in the model so that Equation 14.16 is satisfied. The model could be constructed with bolt fittings at the joints so that the normal pressure could be adjusted.

14.2.2 Material and Fluid Damping

For material damping, the problem is not as direct, so we must resort to some assumptions regarding the response of a structure to random loading. Take the case in which the material and air damping can be completely described by a viscous damping coefficient. For practical purposes, the power spectral density of the displacement of a structure under random loading can be written as (see Equation 11.64)

$$(P.S.)_d = \sum_r \frac{q_r^2}{|Y_r|^2} C_{rr}(\omega) \tag{14.17}$$

where
q is associated with the mode shape of the structure
C_{rr} is the correlation integral that, for all practical purposes, can be considered independent of material damping
Y_r is the response function, which is critically dependent on damping

The response function Y can be written as

$$|Y_r(\omega)| = M_r\sqrt{(\omega_r^2 - \omega^2) + \beta_r^2\omega^2} \tag{14.18}$$

where M_r is the generalized mass for the rth mode, and β_r is the damping coefficient for this mode.

The logarithmic decrement δ_r can be written as

$$\delta_r = \pi\beta_r/\omega_r, \qquad \beta_r = \delta_r\omega_r/\pi = \bar{\beta} \tag{14.19}$$

Fung, Sechler, and Kaplan[10] found that β_r was approximately a constant for various modes of aluminum cylindrical shells tested in air. Thus, at resonance, where $\omega_r = \omega$,

$$Y_r(\omega) = M_r\beta_r\omega_r \tag{14.20}$$

In a structure subjected to random loading, the most severe response in the spectrum will occur at the frequencies corresponding to resonances of the modes. Therefore, at these frequencies,

$$(P.S.)_d \sim \frac{1}{\beta_r^2} = \xi^2 \tag{14.21}$$

Thus, for a model of different material than the full scale,

$$(P.S.)'_d/(P.S.)_d \sim \frac{1/\beta_r'^2}{1/\beta_r^2} = \xi'^2/\xi^2 \tag{14.22}$$

Thus, in order to scale material damping properly at the critical frequencies:

$$\begin{aligned}&\text{Displacement:} && (P.S.)d'/(P.S.)_d = (\xi'^2/\xi^2)\lambda\sqrt{\sigma_E/\sigma_{\rho_s}} \\ &\text{Velocity:} && (P.S.)'_v/(P.S.)_v = (\xi'^2/\xi^2)1/\lambda(\sigma_E/\sigma_{\rho_s})^{3/2} \\ &\text{Acceleration:} && (P.S.)'_a/(P.S.)_a = (\xi'^2/\xi^2)1/\lambda^3(\sigma_E/\sigma_{\rho_s})\end{aligned} \quad (14.23)$$

The ratio ξ'/ξ must be determined experimentally for the different materials. However, a relatively straightforward experiment might be used to do this; possibly an experiment such as the one performed by Fung et al.[10] could be performed on cylindrical shell models made of the prototype and model materials.

A more basic assessment of material damping and its dependency on level of excitation is offered by Crandall.[11]

14.3 More General Formulation of Modeling Laws

In the previous section, the modeling laws for the structure were formulated in an approximate fashion considering material and fluid damping as lumped parameters in an effort to derive some practical modeling relationships. Consider now a more exact formulation by assuming that material damping arises out of a thermal relaxation phenomenon (first discussed by Zener,[12] later generalized by Mason[13] and Biot,[14] and considered more recently by Lazan and Goodman[15]). Furthermore, consider that the vibrating structure is moving in a viscous heat-conducting fluid so that fluid damping may arise out of radiation and viscous losses.

The equation of motion of the solid[16] (x component with body forces) is

$$\frac{\partial^2 u}{\partial x^2}+\frac{\partial^2 u}{\partial y^2}+\frac{\partial^2 u}{\partial z^2}+\frac{1}{1-2\nu}\frac{\partial}{\partial x}\left(\frac{\partial u}{\partial x}+\frac{\partial v}{\partial y}+\frac{\partial \omega}{\partial z}\right)-\rho_s\frac{2(1+\nu)}{E}\frac{\partial^2 u}{\partial t^2} + \frac{2(1+\nu)}{E}X = \frac{2(1+\nu)}{1-2\nu}\frac{\partial(\alpha T)}{\partial x} \quad (14.24)$$

where X = body force per unit volume.

The heat conduction equation of the solid[14] (which assumes material damping by the mechanism of the thermal relaxation) is

$$K_s\left(\frac{\partial^2 T}{\partial x^2}+\frac{\partial^2 T}{\partial y^2}+\frac{\partial^2 T}{\partial z^2}\right) = C_s\frac{\partial T}{\partial t}+T-\frac{\alpha E}{1-2\nu}\frac{\partial}{\partial t}\left(\frac{\partial u}{\partial x}+\frac{\partial v}{\partial y}+\frac{\partial w}{\partial z}\right) \quad (14.25)$$

The equations governing behavior of the fluid are the following:

1. The equation of continuity[17]

$$\frac{\partial \rho_f}{\partial t} + \frac{\partial(\rho_f \dot{u})}{\partial x} + \frac{\partial(\rho_f \dot{v})}{\partial y} + \frac{\partial(\rho_f \dot{w})}{\partial z} = 0 \tag{14.26}$$

2. The Navier–Stokes equations of motion[17] (X component with body forces)

$$\rho_f\left(\frac{\partial \dot{u}}{\partial t} + \dot{u}\frac{\partial \dot{u}}{\partial x} + \dot{v}\frac{\partial \dot{u}}{\partial y} + \dot{w}\frac{\partial \dot{u}}{\partial z}\right) = \rho_f F_x - \frac{\partial p}{\partial x} + \frac{\partial}{\partial \alpha}\left[2\mu\frac{\partial \dot{u}}{\partial x} - \frac{2}{3}\mu\left(\frac{\partial \dot{u}}{\partial x} + \frac{\partial \dot{v}}{\partial y} + \frac{\partial \dot{w}}{\partial z}\right)\right]$$
$$+ \frac{\partial}{\partial y}\left[\mu\left(\frac{\partial \dot{u}}{\partial y} + \frac{\partial \dot{v}}{\partial x}\right)\right] + \frac{\partial}{\partial z}\left[\mu\left(\frac{\partial \dot{u}}{\partial z} + \frac{\partial \dot{w}}{\partial x}\right)\right] \tag{14.27}$$

where F_x = body force per unit mass

3. The equation of heat transfer in the fluid[17,18]

$$\rho_f C_{\rho_f}\left(\frac{\partial T}{\partial t} + \dot{u}\frac{\partial T}{\partial x} + \dot{v}\frac{\partial T}{\partial y} + \dot{w}\frac{\partial T}{\partial z}\right) - \left(\frac{\partial p}{\partial t} + \dot{u}\frac{\partial p}{\partial x} + \dot{v}\frac{\partial p}{\partial y} + \dot{w}\frac{\partial p}{\partial z}\right)$$
$$= k_f\left(\frac{\partial^2 T}{\partial x^2} + \frac{\partial^2 T}{\partial y^2} + \frac{\partial^2 T}{\partial z^2}\right) + \mu\left[2\left(\frac{\partial \dot{u}}{\partial x}\right)^2 + 2\left(\frac{\partial \dot{v}}{\partial y}\right)^2 + 2\left(\frac{\partial \dot{w}}{\partial z}\right)^2 + \left(\frac{\partial \dot{u}}{\partial y} + \frac{\partial \dot{v}}{\partial x}\right)^2\right.$$
$$\left. + \left(\frac{\partial \dot{v}}{\partial z} + \frac{\partial \dot{w}}{\partial y}\right)^2 + \left(\frac{\partial \dot{w}}{\partial x} + \frac{\partial \dot{u}}{\partial z}\right)^2 - \frac{2}{3}\left(\frac{\partial \dot{u}}{\partial x} + \frac{\partial \dot{v}}{\partial y} + \frac{\partial \dot{w}}{\partial z}\right)^2\right] \tag{14.28}$$

The main boundary conditions between fluid and solid to consider for modeling are the following:

$$(T)_{\text{fluid}} = (T)_{\text{solid}} \quad \text{at fluid–solid interface}$$
$$\left(k_f\frac{\partial T}{\partial n}\right)_{\text{fluid}} = \left(K_s\frac{\partial T}{\partial n}\right)_{\text{solid}} \quad \text{at fluid–solid interface} \tag{14.29}$$

$(\text{Displacement})_{\text{fluid}} = (\text{Displacement})_{\text{solid}}$ at fluid–solid interface

$(\text{Normal velocity})_{\text{fluid}} = (\text{Normal velocity})_{\text{solid}}$ at fluid–solid interface

$(\text{Stress})_{\text{fluid}} = (\text{stress})_{\text{solid}}$ at fluid–solid interface

If modeling solids and fluids that are different from the prototype are chosen, then the following scaling parameters must hold:

For the solid:

$$\ell' = \ell\lambda, \quad t' = t\tau, \quad k' = k\chi, \quad T' = T\theta$$

$$E'/E = \sigma_E,\ \rho_s'/\rho_s = \sigma_{\rho_s},\ \mathcal{K}_s'/\mathcal{K}_s = \sigma\mathcal{K}_s,\ \alpha'/\alpha = \sigma_\alpha,\ C_s'/C_s = \sigma_{c_{vs}},\ X'/X = \sigma_X \tag{14.30}$$

For the fluid:

$$\rho_f'/\rho_f = \sigma_{\rho_f}, \mu'/\mu = \sigma_\mu, C_{pf}'/C_{p_f} = \sigma_{C_{p_f}}, k_f'/k_f = \sigma_{k_f}, p_f'/p_f = \sigma_{p_f}, F_x'/F_x = \sigma_{F_x} \tag{14.31}$$

These relations are the fundamental parameters once the model material and model fluid are chosen.

Equations 14.24 and 14.25 can then be written for the solid model as (see Equation 14.9 and again assume the same in model and prototype):

$$\frac{\sigma_E E}{2(1+\nu)}\frac{1}{\lambda}\left[\frac{\partial^2 u}{\partial x^2}+\frac{\partial^2 u}{\partial y^2}+\frac{\partial^2 u}{\partial z^2}+\frac{1}{1-2\nu}\left(\frac{\partial^2 u}{\partial x^2}+\frac{\partial^2 v}{\partial xy}+\frac{\partial^2 w}{\partial xyz}\right)\right]+\sigma_x X$$
$$= \sigma_{\rho_s}\rho_s\frac{\lambda}{\tau^2}\frac{\partial^2 u}{\partial t^2}+\frac{\sigma_E E}{1-2\nu}\sigma_\alpha\alpha\frac{\theta}{\lambda}\frac{\partial T}{\partial \alpha} \tag{14.32}$$

$$\frac{\sigma_{\mathcal{K}_s}\mathcal{K}_s\theta}{\lambda^2}\left(\frac{\partial^2 T}{\partial x^2}+\frac{\partial^2 T}{\partial y^2}+\frac{\partial^2 T}{\partial z^2}\right)=\sigma_{C_s}C_s\frac{\theta}{\tau}\frac{\partial T}{\partial t}+\theta T_r\frac{\sigma_\alpha\alpha\sigma_E E}{1-2\nu}\frac{1}{\tau}\frac{\partial}{\partial t}\left(\frac{\partial u}{\partial x}+\frac{\partial v}{\partial y}+\frac{\partial w}{\partial z}\right) \tag{14.33}$$

Thus, the following relations must hold among the scaling factors in order that the basic differential equations will be the same for model and prototype:

$$\sigma_E/\lambda = \sigma_A\lambda/\tau^2 = \sigma_E\sigma_\alpha\theta/\lambda = \sigma_x \tag{14.34}$$

$$\sigma_{\mathcal{K}_s}/\lambda^2 = \sigma_{c_s}/\tau = \sigma_\alpha\sigma_E/\tau \tag{14.35}$$

The relation $\sigma_E/\lambda = \sigma_{\rho_s}\lambda/\tau^2$ leads to Equation 12.10. The additional relation $\sigma_E/\lambda = \sigma_E\sigma_\alpha\theta/\lambda$ leads to

$$\theta = 1/\sigma_\alpha \tag{14.36}$$

This says that the temperature of the model will be scaled by $1/\sigma_\alpha$. From Equation 12.35, we have the relations

$$\sigma_{K_s}/\lambda^2 = \sigma_{C_s}/\tau \qquad \sigma_{C_s} = \sigma_\alpha\sigma_E \tag{14.37}$$

It is plainly seen that if the same materials are used in model and prototype and testing is done in the same fluids, the first of the equations in Equation 14.37 is inconsistent with the elastic scaling law (Equation 14.34). In fact, the only way that thermoelastic damping can be scaled is by choosing the length scale and the material for the model in such a manner that the coefficient of expansion, heat transfer coefficient, and specific heat of the model and prototype obey Equation 14.37. This does not seem to offer much hope. The only hope for obtaining sensible results for material damping scaling seems to be by use of a relation similar to Equation 14.23. In cases of built-up structures containing many connections, the joint or interface damping will undoubtedly overshadow the material damping, so this inconsistency will play a minor role.

Now consider the fluid. If the equations of the model fluid are written in a similar manner to that for the solid, we obtain

$$\frac{\sigma_{\rho_f}}{\tau}\frac{\partial \rho_f}{\partial t} + \sigma_{\rho_f}\frac{\lambda}{\lambda\tau}\left[\frac{\partial}{\partial x}(\rho_f \dot{u}) + \frac{\partial}{\partial y}(\rho_f \dot{v}) + \frac{\partial}{\partial z}(\rho_f \dot{w})\right] = 0 \tag{14.38}$$

$$\sigma_{\rho_f}\left(\frac{\lambda}{\tau^2}\right)\begin{pmatrix}\text{Left side}\\ \text{of (12.27)}\end{pmatrix} = \sigma_{\rho_f}\sigma_{F_\alpha}\rho_f F_x - \frac{\sigma_{p_f}}{\lambda} + \frac{\partial p}{\partial x} + \frac{\sigma\mu}{\lambda\tau}\left\{\left[2\mu\frac{\partial \dot{u}}{\partial x} + \cdots\right]\right\} \tag{14.39}$$

$$\sigma_{\rho_f}\sigma_{C_{\rho_f}}\rho_f C_{\rho_f}\frac{\sigma_\theta}{\tau}\left(\frac{\partial T}{\partial t} + \cdots\right) - \frac{\sigma_{p_f}}{\tau}\left(\frac{\partial p}{\partial t} + \cdots\right) = \sigma k_f k_f \frac{\theta}{\lambda^2}\left(\frac{\partial^2 T}{\partial x^2} + \cdots\right) + \frac{\sigma_\mu \mu}{\tau^2}\left(2\frac{\partial \dot{u}}{\partial x} + \cdots\right) \tag{14.40}$$

The basic scaling relations that must hold are therefore

$$\sigma_{\rho_f}\lambda/\tau^2 = \sigma_{p_f}/\lambda = \sigma_\mu/\lambda\tau = \sigma_{\rho_f}\sigma_{F_x} \tag{14.41}$$

$$\sigma_{\rho_f}\sigma_{c_{\rho_f}}\sigma\theta/\tau = \sigma_{\rho_f}/\tau = \sigma_{k_f}\theta/\lambda^2 = \sigma_\mu/\tau^2 \tag{14.42}$$

The relation $\sigma_{\rho_f}\lambda/\tau^2 = \sigma_\mu/\lambda\tau$ leads to the constant Reynold's number scaling:

$$\rho'_f v'\ell'/\mu' = \rho v\ell/\mu \tag{14.43}$$

(noting that $\sigma_v = \lambda/\tau$).

This relation inherently contains the viscous damping offered by the fluid to the solid, noting the fluid–solid boundary conditions and introducing

$$\frac{\sigma_{\rho_f}}{\sqrt{\sigma_{\rho_s}/\sigma_E}} = \sigma_\mu/\lambda \tag{14.44}$$

If the same fluids and solids are used for the model and prototype, this will lead again to an inconsistency in Equation 14.44 if $\lambda \neq 1$. The only way that we can scale fluid viscous damping is to choose a length scale and model materials that obey Equation 14.44. The second scaling relation from Equation 14.44 is

$$\sigma_{\rho f}\lambda/\tau^2 = \sigma_{\rho f}/\lambda$$

$$\sigma_{\rho f}\sigma_{\upsilon^2}/\sigma_{pf} = 1 \quad \text{or} \quad \frac{\rho'_f}{\rho_f}\frac{v'^2}{v^2}\Big/p'/p \quad \text{or} \quad \frac{P'}{\rho'_f v^2\ell'^2} = \frac{P}{\rho_f v^2\ell'^2} \tag{14.45}$$

where p denotes pressure and P denotes force.

This is Newton's universal similitude law,[19] which must hold between the model- and full-scale fluids. It inherently contains the scaling relations for mass loading and radiation damping offered by the fluid to the solid. Now, going back to the boundary condition between solid and fluid, it is seen that stresses and pressures must be scaled by σ_E. Thus,

$$\sigma_{\rho_f}\lambda/\tau^2 = \sigma_E/\lambda \tag{14.46}$$

Noting that $\lambda^2/\tau^2 = \sigma_E/\sigma_{\rho_s}$ from elastic scaling, we finally obtain

$$\sigma_{\rho_f} = \sigma_{\rho_s} \tag{14.47}$$

This says that the scale factors for the density of the fluids and solids for model and prototype must be the same in order for mass loading and radiation damping to scale properly. If the same density fluids are used for the model testing, then the same density solids must be used.

The fluid heat transfer equation gives scaling parameters that must satisfy the following:

$$\begin{aligned}
&\sigma_{\rho_f}\,\sigma_{c_{\rho_f}}\,\theta = \sigma_{\rho_f} \quad \text{or} \quad \sigma_{\rho_f}\sigma_{c_{\rho_f}}\theta = \sigma_E \\
&\sigma_{\rho_f}\,\sigma_{c_{\rho_f}}/\tau = \sigma_{k_f}/\lambda^2 \quad \text{or} \quad \sigma_{\rho_f}\sigma_{c_{\rho_f}} = \frac{\sigma_{k_f}}{\lambda}\sqrt{\sigma_A/\sigma_E} \\
&\sigma_{\rho_f}\,\sigma_{c_{\rho_f}}\,\theta = \sigma_\mu/\tau \quad \text{or} \quad \sigma_{\rho f}\,\sigma_{c_{\rho_f}}\theta = \frac{\sigma\mu}{\lambda}\sqrt{\sigma_{\rho_s}/\sigma_{\rho s}}
\end{aligned} \tag{14.48}$$

The same inconsistency between these relations and the elastic scaling relation is obtained when the same materials are used for model and full scale. In cases where heating of the fluid is not a primary problem, the heat transfer characteristics of the fluid can be neglected completely, and it is only necessary to consider the fluid–solid problem without Equation 14.28.

14.4 Some General Considerations in Modeling

The use of very small-scale models of material different from that of the prototype offers a very efficient and inexpensive way to obtain order of magnitude answers that would ordinarily be very difficult or impossible to obtain. This type of scaling also offers an opportunity to use much lower pressure excitation levels than the full scale. There might also be a possibility of using lightweight, easily constructed plastic models to perform initial tests in the same manner as those performed by Sankey and Wright,[20] but with a more complete assessment of damping.

Invariably, damping offers a big problem in using scale model results to extrapolate full-scale response. If we limit our discussion to the resonance response, then damping could be introduced approximately in the scaling as shown in Equation 14.21 if ξ were constant for a given material. This resonant limitation is not serious in random loading since the major response at each frequency of interest is usually composed of primarily resonant contributions in modes close to this frequency.

As shown in the previous section in Equation 14.37, material damping of the thermoelastic type cannot easily be scaled—nor can viscous air damping arising from air viscosity, unless tests are conducted in "thinner" air or in different fluids. The main hope with model testing built-up structures under random loading is that joint or interfacial slip will be the main source of damping both in the model and in the full scale, whether they are of the same or different materials. If this were true and if frictional stresses were scaled properly according to Equation 14.16, then Relations 14.13 and 14.14 will hold for the random response. If one uses model materials with high material damping, such as plastics, one must be sure that the joint damping is modeled properly before resorting to approximate equations of the form of Equation 14.23. Equation 14.23 must be used with extreme caution since it will only hold at resonant frequencies and will hold only if ξ can be shown to be a function of material only.

References

1. Greenspon, J. E. 1964. Modeling of spacecraft under random loading. JG Engineering Research Associates, NASA CR–132.
2. Weber, W. 1949. The universal principle of similitude in physics and its relation to the dimensional theory and the science of models. David Taylor Model Basin trans. 200, Sept., 1949 (DTMB is now Naval Ship Res. & Dev. Center, Washington, DC).
3. Love, A. E. H. 1944. *A treatise on the mathematical theory of elasticity*, 293. New York: Dover Publications.
4. Junger, M. C. The scaling laws governing the dynamic response of a structure to a turbulent boundary layer. *U.S. Navy Journal of Underwater Acoustics* April: 439.
5. Junger, M. C. The scaling laws governing the dynamic response of a structure to a turbulent boundary layer. *U.S. Navy Journal of Underwater Acoustics* April: 439.
6. Wright, D. V., Miller, D. F., Akey, J. G., and Hogg, A. C. 1962. Vibration transmission and impedance of basic foundation structures. Westinghouse Research Labs, BuShips contract Nobs 72326.
7. Mixson, J. S., Catherine, J. J., and Arman, A. 1963. Investigation of the lateral vibration characteristics of a 1/5 scale model of Saturn SA–1. NASA TN 1593.
8. Baker, W. E., ed. 1963. Use of models and scaling in shock and vibration. ASME Colloquium, Nov., 1963 (published under auspices of the American Society of Mechanical Engineers).
9. Calligeros, J. M., and Dugundji, J. 1959. Similarity laws required for experimental aerothermoelastic studies. Aerolastic and Structures Research Laboratory, M.I.T., tech. rep. 75–1.
10. Fung, Y. C., Sechler, E. E., and Kaplan, A. 1960. On the vibration of cylindrical shells under internal pressure. *Journal of Aeronautical Science* 24: 650–660.

11. Crandall, S. H. 1962. On scaling laws for material damping. NASA TN D–1467, Dec., 1962.
12. Zener, C. 1948. *Elasticity and anelasticity of metals,* chap. VII. Chicago, IL: University of Chicago Press.
13. Mason, W. P. 1949. *Piezoelectric crystals and their applications to ultrasonics,* chap. 3 and appendix. New York: D. Van Nostrand Co., Inc.
14. Biot, M. A. 1956. Thermoelasticity and irreversible thermodynamics. *Journal of Applied Physics* 27 (3): 240.
15. Lazan, B. J., and Goodman, L. E. 1961. Material interface damping. In *Shock and vibration handbook,* vol. 2, ed. C. M. Harris and C. E. Crede. New York: McGraw-Hill.
16. Parkus, H. 1962. Thermal stresses. In *Handbook of engineering mechanics,* ed. W. Flugge. New York: McGraw–Hill.
17. Brown, C. E. 1962. Fluid mechanics—Basic concepts. In *Handbook of engineering mechanics,* ed. W. Flugge. New York: McGraw–Hill.
18. Kay, J. M. 1963. *An introduction to fluid mechanics and heat transfer,* 198. Cambridge, England: Cambridge University Press.
19. Weber, W. 1949. The universal principle of similitude in physics and its relation to the dimensional theory and the science of models. David Taylor Model Basin trans. 200, Sept., 1949 (DTMB is now Naval Ship Res. & Dev. Center, Washington, DC).
20. Sankey, G. O., and Wright, D. V. 1963. Vibration characteristics of propulsion machinery subbases as determined by plastic model tests. Westinghouse Research Labs, BuShips contract Nobs 86809.

Index

D

P

Q

R